AUSTRALIAN MATHEMATICAL SOCIETY LECTURE SERIES

Editor-in-Chief: Dr. S.A. Morris, Department of Mathematics, Statistics and Computing Science, University of New England, Armidale, N.S.W. 2351, Australia

Subject Editors:
Professor C.J. Thompson, Department of Mathematics, University of Melbourne, Parkville, Victoria 3052, Australia
Professor C.C. Heyde, Department of Statistics, University of Melbourne, Parkville, Victoria 3052, Australia
Professor J.H. Loxton, Department of Pure Mathematics, University of New South Wales, Kensington, New South Wales 2033, Australia

1 Introduction to Linear and Convex Programming, N. CAMERON
2 Manifolds and Mechanics, A. JONES, A. GRAY & R. HUTTON
3 Introduction to the Analysis of Metric Spaces, J.R. GILES
4 An Introduction to Mathematical Physiology and Biology, J. MAZUMDAR

D1106184

Australian Mathematical Society Lecture Series. 5

2-Knots and their Groups

Jonathan Hillman
Macquarie University,
Australia

The right of the University of Cambridge to print and sell all manner of books was granted by Henry VIII in 1534. The University has printed and published continuously since 1584.

CAMBRIDGE UNIVERSITY PRESS

Cambridge

New York New Rochelle Melbourne Sydney

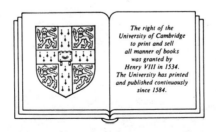

Published by the Press Syndicate of the University of Cambridge
The Pitt Building, Trumpington Street, Cambridge CB2 1RP
32 East 57th Street, New York, NY 10022, USA
10, Stamford Road, Oakleigh, Melbourne 3166, Australia

First Published 1989

Printed in Great Britain at the University Press, Cambridge

Library of Congress cataloging in publication data available

British Library cataloguing in publication data available

ISBN 0 521 37812 5 paperback

CONTENTS

Pour les noeuds de S^2 en S^4 on ne sait pas grand chose

Preface

Since Gramain wrote the above words in a Seminaire Bourbaki report on classical knot theory in 1976 there have been major advances in 4-dimensional topology, by Casson, Freedman and Quinn. Although a complete classification of 2-knots is not yet in sight, it now seems plausible to expect a characterization of knots in some significant classes in terms of invariants related to the knot group. Thus the subsidiary problem of characterizing 2-knot groups is an essential part of any attempt to classify 2-knots, and it is the principal topic of this book, which is largely algebraic in tone. However we also draw upon 3-manifold theory (for the construction of many examples) and 4-dimensional surgery (to establish uniqueness of knots with given invariants). It is the interplay between algebra and 3- and 4-dimensional topology that makes the study of 2-knots of particular interest.

Kervaire gave homological conditions which characterize high dimensional knot groups and which 2-knot groups must satisfy, and showed that any high dimensional knot group with a presentation of deficiency 1 is a 2-knot group. Bridging the gap between the homological and combinatorial conditions appears to be a delicate task. For much of this book we shall make a further algebraic assumption, namely that the group have an abelian normal subgroup of rank at least 1. This is satisfied by the groups of many fibred 2-knots, including all spun torus knots and cyclic branched covers of twist spun knots. The evidence suggests that if the abelian subgroup has rank at least 2 then the group is among these, and the problem is then related to that of characterizing 3-manifold groups and their automorphisms. Most known knots with such groups can be characterized algebraically, modulo the s-cobordism theorem. However in the rank 1 case there are examples which are not the groups of fibred knots, and here less is known.

The other class of groups that is of particular interest as it contains the groups of (spun) classical knots consists of those which have cohomological dimension 2 and deficiency 1. (If some standard conjectures hold these conditions are equivalent for knot groups). One striking member of this class is the group Φ with presentation $\langle a,t \mid tat^{-1} = a^2\rangle$, whose commutator subgroup is a torsion free rank 1 abelian group. All other knot groups with deficiency 1 and nontrivial torsion free abelian normal subgroups are iterated free products of torus knot groups, amalgamated over central copies of Z, and are the groups of fibred 2-knots. We show that any knot with such a group (and more generally, whose group has free commutator subgroup) can be characterized algebraically, modulo the s-cobordism theorem. Together these two classes contain the groups of the most familiar and important examples of 2-knots. However we have by no means completed their classification, and the problem of organizing the groups outside these classes remains quite open. (The formation of sums and satellites should play a part here).

We shall now outline the chapters in somewhat greater detail. In Chapter 1 we give the basic definitions and background results on the geometry of knots and we show how the classification of higher dimensional knots can be reduced (essentially) to the classification of the closed manifolds built from the ambient spheres by surgery on such knots. As far as possible these definitions and results have been formulated so as to apply in all dimensions. We have chosen to work in the TOP category as our chief interest is in the 4-dimensional case, where PL or (equivalently) DIFF techniques are not yet adequate.

In Chapter 2 we give Kervaire's characterization of high dimensional knot groups, and variations on this theme: link groups, commutator subgroups of knot groups, centres of knot groups. We also give his partial results on 2-knot groups. Counter examples to show that not all high dimensional knot groups can be 2-knot groups were found independently by various people; most of their arguments used duality in the infinite cyclic cover of the exterior of the knot. We review some of these arguments, and we show that the exterior of a nontrivial n-knot with $n > 1$ is never aspherical, giving Eckmann's proof via duality in the universal cover.

Chapter 3 contains our key result. We show that in contrast to the theorem of Dyer-Vasquez and Eckmann just quoted the closed 4-manifold obtained by surgery on a 2-knot is often aspherical. If T is the maximal locally-finite normal subgroup of a 2-knot group π and π/T has an abelian normal subgroup of rank 1 such that the quotient has finitely many ends and if a further, technical condition (that may prove to be redundant) holds then either π' is finite or $\pi/T = \Phi$ or π/T is an orientable Poincaré duality group over Q of formal dimension 4. The latter is also true if π/T has an abelian normal subgroup of rank greater than 1.

In the next three chapters we examine these cases separately. In Chapter 4 we determine the 2-knot groups with finite commutator subgroup. All of these can be realized by fibred 2-knots, and many by twist spun classical knots. We show also that if $\pi/T = \Phi$ and T is nontrivial then it must be infinite; in fact we believe that in this case T must be trivial. In Chapters 5 and 6 we consider the Poincaré duality cases. Here there is a further subdivision of cases, according to the rank of the abelian normal subgroup (which must be at most 4). All the known examples with a torsion free abelian normal subgroup of rank 2 derive from twist spun torus knots. The groups of aspherical Seifert fibred 3-manifolds may be characterized as PD_3-groups which have subgroups of finite index with nontrivial centre and infinite abelianization. Using this, we give an algebraic characterization of the groups of 2-knots which are cyclic branched covers of twist spins of torus knots.

In Chapter 6 we determine the 2-knot groups with abelian normal subgroups of rank greater than 2, and the results of these three chapters are combined to show that if π has an ascending series whose factors are locally-finite or locally-nilpotent, then it is in fact locally-finite by solvable. If moreover π has an abelian normal subgroup of positive rank then it is finite by solvable, and we describe all such groups. (We doubt that there are any other 2-knot groups with such ascending series).

In the last two chapters we attempt to recover 2-knots from group theoretic invariants. As we observe in Chapter 1 a knot K is determined up to changes of orientation and "Gluck reconstruction" by a certain closed 4-manifold $M(K)$ together with a conjugacy class in the knot group $\pi_1(M)$. We first try to determine the homotopy type of M in terms of algebraic invariants. The problem of the homeomorphism type may then be reduced to standard questions of surgery. For knots whose group is torsion free and polycyclic we are completely successful, for the surgery techniques are then available to solve the problem. We show also that if the commutator subgroup is an infinite, nonabelian nilpotent group then, excepting for two such groups, the knot is determined up to inversion by its group alone.

Freedman has shown that surgery techniques apply whenever the group is as in Chapter 6, but in general it is difficult to compute the obstructions. On the other hand, for many fibred 2-knots we can determine the (simple) homotopy type and show that the surgery obstructions are 0, but it is not yet known whether 5-dimensional s-cobordisms with such groups are always products.

After Chapter 8 there are two appendices. The first considers the 4-dimensional geometries that can be supported by some $M(K)$. (Among these are some complex surfaces). In the second it is shown that certain Cappell-Shaneson 2-knots are reflexive if and only if every totally positive unit in the cubic number field generated by a root of the Alexander polynomial is a square in that field. After there is a list of open questions on 2-knots and related topics. Some of these are well known and very difficult; others are more technical and algebraic, but have also resisted solution so far.

As the algebra used in this book may perhaps be unfamiliar for many topologists we would like to stress here that our principal references have been the Queen Mary College lecture notes of Bieri for homological group theory, and the text of Robinson for other aspects of group theory.

I would like to thank William Dunbar, Ross Geoghegan, John Groves, Laci Kovács, Peter Kropholler, Darryl McCullough, Mike Mihalik, Peter M. Neumann, Steve Plotnick, Peter Scott and Shmuel Weinberger for their correspondance and advice on various aspects of this work. I would also like to acknowledge the support of the U.K. Science and Engineering Research Council (as a Visiting Fellow at the University of Durham), which enabled me to meet Peter Kropholler and Peter Scott, and of the Australian Research Grants Scheme, for a grant which brought Steve Plotnick to Macquarie University in mid-1987.

Macquarie University

Chapter 1 KNOTS AND RELATED MANIFOLDS

In this chapter are the basic definitions and constructions of the objects that we shall study. In particular we show how the classification of higher dimensional knots can be reduced (essentially) to the classification of certain closed manifolds. We also give a number of results on the geometry of these objects, for the most part without proof, as we shall not use them in a crucial way in our arguments later (which are primarily algebraic). We shall first state some of our conventions on notation and terminology.

Let $D^n = \{ \langle x_1,...x_n \rangle \ in \ R^n \mid \Sigma x_i^2 \leqslant 1 \}$ be the n-disc and let $S^n = \partial D^{n+1}$ be the n-sphere. The standard orientation of R^n induces an orientation of D^n, and hence one of S^{n-1} by the convention that the boundary of an oriented manifold be oriented compatibly with taking the inward normal last (cf. [RS: pages 44-45]). We shall always assume that these standard discs and spheres have been given such standard orientations.

An *embedding* is a 1-1 map which is a homeomorphism onto its image. All isotopies of locally flat embeddings of manifolds in manifolds shall be *ambient* isotopies. The inclusion of R^{n+1} into R^{n+2} as the hyperplane defined by the equation $x_{n+2} = 0$ induces the *equatorial* embedding $e_n:S^n \to S^{n+1}$. The interior of a subset N of a topological space shall be denoted by $int\, N$.

The expression $A \cong B$ means that the objects A and B are isomorphic in some category appropriate to the context. When there is a canonical isomorphism, or after a particular isomorphism has been chosen, we shall write $A = B$. (For instance the fundamental group of a circle is infinite cyclic, and choosing an isomorphism with Z corresponds to choosing an orientation for the circle). Qualifications shall usually be omitted when there is no risk of ambiguity. In particular, we may abbreviate $X(K)$, $M(K)$ and πK as X, M and π respectively.

Knots

An *n-knot* is a locally flat embedding $K:S^n \to S^{n+2}$. (We shall also use the terms "classical knot" when $n = 1$, "higher dimensional knot" when $n \geqslant 2$ and "high dimensional knot" when $n \geqslant 3$). Such a knot is determined up to isotopy by its image $K(S^n)$, considered as an oriented codimension 2 submanifold of S^{n+2}, and so we may let K also denote this submanifold. Two n-knots K_0 and K_1 are isotopic if (and only if) there is an orientation preserving self homeomorphism h of S^{n+2} such that $hK_0 = K_1$, for such a map h is isotopic to the identity [KS: page 292]. Thus for instance if r_n is an orientation reversing self homeomorphism of S^n the knots K, $rK = r_{n+2}K$, $K\rho = Kr_n$ and $-K = rK\rho$ may all be distinct. (Note that the images of K and $K\rho$ are the same, considered as unoriented submanifolds). The knot K is *invertible*, *+amphicheiral* or *−amphicheiral* if it is isotopic to $K\rho$, rK or $-K$ respectively. An n-knot is *trivial* if it is isotopic to the unknot $e_{n+1}e_n$. By the uniqueness of discs an n-knot is trivial if and only if it bounds a locally flat $(n+1)$-disc in S^{n+2}.

If $n \geqslant 4$ then each n-knot is isotopic to a PL n-knot which is unique up to PL (ambient) isotopy, for then $H^i(S^{n+2},K;Z/2Z) = 0$ for $i = 3$ and 4, so the Kirby–Siebenmann obstructions to existence and uniqueness of PL triangulations vanish [KS: Essay IV.10]. All the examples of 2-knots that we shall construct below shall be PL with respect to some triangulation of S^4. However as it is not yet known whether all triangulations of S^4 are PL equivalent, and as the 4-dimensional surgery techniques that we wish to use to characterize certain 2-knots have only been established in the TOP category (and then only for a restricted class of fundamental groups), the above definition of knot seems most suitable.

The exterior

Every (locally flat) n-knot with $n = 2$ is flat [KS 1975] and so there is an embedding $j:S^n \times D^2 \to S^{n+2}$ onto a closed neighbourhood N of K, such that $j|S^n\times\{0\} = K$ and $\partial N = j(S^n \times S^1)$ is bicollared in S^{n+2}. We may assume that j is orientation preserving, and then it is unique up to isotopy *rel* $S^n\times\{0\}$. These results (and their proofs) remain valid when $n = 2$, by Quinn's solution of the 4-dimensional annulus conjecture, together with surgery over Z (cf. [Qu 1982] and [FQ]).

The *exterior* of K is the compact $(n+2)$-manifold $X(K) = S^{n+2} - int\, N$, which is well defined up to homeomorphism as N is unique up to isotopy *rel* K. It inherits an orientation from S^{n+2}, and has boundary homeomorphic to $S^n \times S^1$. The interior of X is homeomorphic to the knot complement $S^{n+2}-K$. The *knot group* is $\pi K = \pi_1(X(K))$. By general position, every element of πK can be represented by an oriented simple closed curve in X, and if $n \geqslant 2$ each conjugacy class in πK (i.e. free homotopy class of maps from S^1 to X) corresponds to an unique isotopy class of such curves. In particular, any oriented simple closed curve isotopic to the oriented boundary of a transverse disc (i.e. to $\{j(s)\} \times S^1$) is called a *meridian* of K, and we shall also use this term to denote the corresponding elements of πK.

By Alexander duality $H_i(X;Z) \cong Z$ if $i = 0$ or 1 and is 0 otherwise. The meridians are all homologous and generate $H_1(X;Z)$ (by the Mayer–Vietoris sequence for $S^{n+2} = X \cup S^n \times D^2$) and so determine a canonical isomorphism $H_1(X;Z) = Z$.

The exterior of a trivial n-knot is homeomorphic to $D^{n+1} \times S^1$. Conversely if $X(K)$ is homotopy equivalent to S^1 then K is trivial [Pa 1957, St 1963, Le 1965, Sh 1968, Fr 1983]. For if γ is a meridian in the interior of X, it has a product neighbourhood U (as X is orientable) and if $X \sim S^1$ then $X - int\, U$ is an s-cobordism and so a product, provided $n \geqslant 3$. It is then easy to see that K bounds a disc in S^{n+2}. (This argument works also in the PL and DIFF categories). This criterion for triviality is also correct when $n = 1$ [Pa 1957] and $n = 2$ [Fr 1983], although the proofs are different. (Note that there may be PL 2-knots with group Z which are *not* PL isotopic to the unknot). The assumption on X can be weakened considerably (see below); in particular if $n = 1$ or 2 an n-knot is trivial if and only if its group is infinite cyclic.

Gluck showed that when $n = 2$ the group of pseudoisotopy classes of self homeomorphisms of $S^n \times S^1$ is $(Z/2Z)^3$, generated by reflections in either factor and by the map τ given by $\tau(x,y) = (\theta(y)(x),y)$ for all x in S^n and y in S^1, where $\theta{:}S^1 \to SO(n+1)$ is an essential map [Gl 1962]. Browder and Kato extended his result to all higher dimensions [Br 1967, Ka 1969]. As any self homeomorphism of $S^n \times S^1$ extends across $D^{n+1} \times S^1$ (provided $n \geqslant 2$), the closed $(n+2)$-manifold $M(K) = X(K) \cup D^{n+1} \times S^1$ obtained from S^{n+2} by surgery on K is also well defined, and it inherits an orientation from S^{n+2} via X. Since up to homotopy X is the complement of $\{0\}\times S^1$ in M, the inclusion of X into M induces an isomorphism $\pi K = \pi_1(M)$, by general position (for $n \geqslant 2$). The Euler characteristic of M is $\chi(M) = \chi(S^{n+2})-\chi(S^n\times D^2)+\chi(D^{n+1}\times S^1) = 0$. In fact $H_i(M;Z)$ is Z for $i = 0, 1, n+1$ or $n+2$ and is 0 otherwise, as follows easily from the Mayer-Vietoris sequence for $M = X\cup D^{n+1}\times S^1$. (The orientations of S^{n+2} and K determine canonical isomorphisms). We shall in fact study 2-knots K through the corresponding closed oriented 4-manifolds $M(K)$.

There is however an ambiguity when we attempt to recover K from $M(K)$. The cocore $\gamma = \{0\}\times S^1 \subset D^{n+1}\times S^1 \subset M$ of the original surgery is well defined up to isotopy by the conjugacy class of a meridian in $\pi K = \pi_1(M)$. (In fact the orientation of γ is irrelevant for what follows). Its normal bundle is trivial, so γ has a product neighbourhood, P say, and we may assume that $M-int\,P = X$. But there are two essentially distinct ways of identifying ∂X with $S^n\times S^1 = \partial(S^n\times D^2)$, modulo self homeomorphisms of $S^n\times S^1$ that extend across $S^n\times D^2$, provided $n \geqslant 2$. If we reversed the original construction of M we recover $(S^{n+2}, K) = (X\cup_{j\partial}S^n\times D^2, S^n\times\{0\})$. If however we identify $S^n\times S^1$ with X by means of $j\tau$ we obtain a new pair $(\Sigma, K^*) = (X\cup_{j\tau}S^n\times D^2, S^n\times\{0\})$. By van Kampen's theorem Σ is simply connected, and it is easily seen to have the homology of S^{n+2}. Therefore Σ is homeomorphic to S^{n+2}. We may assume that the homeomorphism is orientation preserving. Thus we obtain a new n-knot K^*, which we shall call the *Gluck reconstruction* of K. The knot K is said to be *reflexive* if it is determined as an unoriented submanifold by its exterior, i.e. if K^* is isotopic to K, rK, $K\rho$ or $-K$.

By the work of Browder, Gluck and Kato, if $n \geqslant 2$ there are at most two n-knots (up to change of orientations) with a given exterior, i.e. if there is an orientation preserving homeomorphism from $X(K_1)$ to $X(K)$ then K_1 is isotopic to K, K^*, $K\rho$ or $K^*\rho$. If the homeomorphism also preserves the homology class of the meridians then K_1 is isotopic to K or K^*. (The long-standing conjecture that each classical knot is determined up to orientation by its exterior has finally been confirmed [GL 1988]. The argument involves some of the deepest results of 3-manifold topology). Thus a knot K is determined up to an ambiguity of order at most 2 by $X(K)$, or (if $n \geqslant 2$) equivalently by $M(K)$ together with the conjugacy class of a meridian in πK. Cappell and Shaneson gave the first examples of knots which are not reflexive [CS 1976]. Their method works for each $n \geqslant 2$ provided that certain $(n+1)\times(n+1)$ integral matrices exist; at present such matrices have been found only for $n = 2$, 3, 4 and 5. Gordon gave a different family of examples when $n = 2$, all of which are PL knots with respect to the standard triangulation of S^4 [Go 1976].

Covering spaces and equivariant homology

We shall let $\widetilde{X}(K)$ and $X'(K)$ denote the *universal* and *maximal abelian* covering spaces (respectively) of $X(K)$. Similarly $\widetilde{M}(K)$ and $M'(K)$ are the corresponding covering spaces of $M(K)$. The fundamental group of X' (and of M', provided that $n \geqslant 2$) is the commutator subgroup π' of the knot group, and by the Hurewicz theorem $\pi/\pi' = H_1(X;Z) = Z$. Thus the cover X'/X is also known as the *infinite cyclic* cover of the knot exterior.

The homology and cohomology of such covering spaces are modules over the group ring of the covering group, and satisfy a form of equivariant Poincaré duality. We shall describe this in somewhat greater generality. Let P be a closed orientable m-manifold with fundamental group G. Up to homotopy we may approximate P by a finite cell complex [KS: Essay III.4]. Let H be a normal subgroup of G and let P_H be the corresponding covering space. We may then lift the cellular decomposition of P to an equivariant cellular decomposition of P_H. The cellular chain complex C_* of P_H with coefficients in a commutative ring R is then a complex of left $R[G/H]$-modules with respect to the action of the covering group G/H. Moreover C_* is a complex of free modules, with a finite basis obtained by choosing one lift of each cell of P. The ith *equivariant homology* module of P with coefficients $R[G/H]$ is the left module $H_i(P;R[G/H]) = H_i(C_*)$, which is clearly isomorphic to $H_i(P_H;R)$ as an R-module, with the action of the covering group determining the $R[G/H]$-structure. The ith *equivariant cohomology* module of P with coefficients $R[G/H]$ is the right module $H^i(P;R[G/H]) = H^i(Hom_{R[G/H]}(C_*,R[G/H]))$, which may be interpreted as cohomology of

P_H with compact supports.

If N is a right $R[G]$-module we shall let $\overline{N}$ denote the left module with the same underlying R-module and the conjugate G-action, determined by $g.n = ng^{-1}$ for g in G and n in N.

The equivariant homology and cohomology are related by Poincaré duality isomorphisms $H_i(P;R[G/H]) = \overline{H^{m-i}(P;R[G/H])}$ and by a Universal Coefficient spectral sequence with E_2 term

$$E_2^{pq} = Ext^q_{R[G/H]}(H_p(P;R[G/H]),R[G/H]) \Rightarrow H^{p+q}(P;R[G/H]),$$

in which the differential d_r has bidegree $(1-r,r)$. If J is a normal subgroup of G which contains H there is also a Cartan-Leray spectral sequence relating the homology of P_H to that of P_J, with E^2 term

$$E^2_{pq} = Tor_p^{R[G/H]}(H_q(P;R[G/H]),R[G/J]) \Rightarrow H_{p+q}(P;R[G/J])$$

in which the differential d^r has bidegree $(-r,r-1)$. There are similar definitions and results for manifold pairs $(P,\partial P)$ and for (co)homology with more general coefficients [W]. For more information on these spectral sequences see [McC].

When K is an n-knot, $P = X(K)$ or $M(K)$, $G = \pi K$ and $H = \pi'$, the group ring $Z[\pi/\pi']$ is the ring of integral Laurent polynomials $\Lambda = Z[Z] = Z[t,t^{-1}]$. Since Λ is noetherian the homology and cohomology of a finitely generated free Λ-chain complex are also finitely generated. The augmentation module Z has projective dimension 1 as it has a short free resolution $0 \to \Lambda \to \Lambda \to Z \to 0$. Therefore the Cartan-Leray spectral sequence for the projection of X' onto X (or of M' onto M) reduces to a long exact sequence (the Wang sequence of the map $X' \to X$):

$$\cdots \to H_i(X;\Lambda) \to H_i(X;\Lambda) \to H_i(X;Z) \to H_{i-1}(X;\Lambda) \to \cdots.$$

Since X has the homology of a circle, it follows that all the maps $t-1{:}H_i(X;\Lambda) \to H_i(X;\Lambda)$ are surjective for $i > 0$. Therefore they are bijective (since the modules are noetherian) and so the homology modules are all torsion Λ-modules. In particular $Hom_\Lambda(H_p(X;\Lambda),\Lambda) = 0$ for all p so the Universal Coefficient spectral sequence collapses to a collection of short exact sequences

$$0 \to Ext^2_\Lambda(H_{p-2}(X;\Lambda),\Lambda) \to H^p(X;\Lambda) \to Ext^1_\Lambda(H_{p-1}(X;\Lambda),\Lambda) \to 0.$$

(There are very similar results for $H_*(M;\Lambda)$).

The infinite cyclic covering spaces X' and M' behave homologically much like $(n+1)$-manifolds with boundary S^n and empty respectively, at least if we use field coefficients [Mi 1968, Ba 1980]. If $H_i(X;\Lambda) = 0$ for $1 \leqslant i \leqslant (n+1)/2$ then X' is acyclic; thus if also $\pi = Z$ then X is homotopy equivalent to S^1 and so K is trivial. All the classifications of higher dimensional knots to date assume that the knot group is Z and that the infinite cyclic cover of the exterior is highly connected. An n-knot K is called *r-simple* if $X'(K)$ is r-connected; if $n \geqslant 3$ and the knot is $[(n-1)/2]$-simple it is *simple*. (The word is used in a different sense in connection with classical knots). Levine classified simple $(2q+1)$-knots with $q \geqslant 2$ by means of Seifert matrices [Le 1970]; this was reformulated (and reproven) in terms of the duality pairing on the middle dimensional homology by Kearton [Ke 1975]. After Kearton and Kojima had each classified some significant subclasses of simple $2q$-knots with $q \geqslant 4$ by means of analogous (but more complicated) pairings, Farber finished this task as an application of his classification of *stable* n-knots (i.e. those which are r-simple for some $r \geqslant (n+1)/3$) in terms of stable homotopy pairings [Fa 1983]. By analogy with results on the rational homotopy type of highly connected manifolds, one might hope that an r-simple n-knot K in the "formal range" with $r \geqslant (n-1)/4$ would be determined up to finite ambiguity by the cohomology algebra $H^*(M(K);\Lambda)$. (Ideally one might hope to classify all 1-simple knots up to finite ambiguity by algebraic invariants, but it is not yet clear what these should be).

When $n = 1$ or 2 it is more profitable to work with the universal cover $\widetilde{X}$ (or $\widetilde{M}$). When $n = 1$ the universal cover of X is contractible; this is a remarkable result of Papakyriakopoulos [Pa 1957]. In higher dimensions X is aspherical only when the knot is trivial, as we shall see in Chapter 2. Nevertheless under rather mild assumptions on the group of a 2-knot K the closed 4-manifold $M(K)$ *is* aspherical. (See Chapter 3). This is the main reason that we choose to work with $M(K)$ rather than $X(K)$.

Knot sums, factorization and satellites

The *sum* of two n-knots K_1 and K_2 may be defined (up to isotopy) as the n-knot $K_1\#K_2$ obtained as follows. Let $D^n(\pm)$ denote the upper and lower hemispheres of S^n. We may isotope K_1 and K_2 so that each $K_i(D^n(\pm))$ is contained in $D^{n+2}(\pm)$, $K_1(D^n(+))$ is a trivial n-disc in $D^{n+2}(+)$, $K_2(D^n(-))$ is a trivial n-disc in $D^{n+2}(-)$ and $K_1|S^{n-1} = K_2|S^{n-1}$ (as the oriented boundaries of the images of $D^n(-)$). Then we let $K_1\#K_2 = K_1|D^n(-)\cup K_2|D^n(+)$. By van Kampen's theorem (used several times) $\pi(K_1\#K_2) = \pi K_1 *_Z \pi K_2$ where the amalgamating subgroup is generated by a meridian in each knot group. It is not hard to see that $X'(K_1\#K_2)$ is homotopy equivalent to $X'(K_1)\vee X'(K_2)$ and so in particular $\pi'(K_1\#K_2) = \pi'(K_1)*\pi'(K_2)$.

When $n = 1$ this construction corresponds to tying two knots consecutively in the same loop. We say that a knot is *irreducible* if it is not the sum of two nontrivial knots. Schubert showed that each 1-knot has an essentially unique factorization as a sum of irreducible knots (and so irreducible 1-knots are usually called *prime* knots) [Sch 1949]. In higher dimensions this is false in general. However Dunwoody and Fenn have shown that every n-knot with $n \geqslant 3$ admits some finite factorization into irreducible knots, and moreover for each knot there is a bound on the length of such factorizations [DF 1987]. As the only geometric result needed for their argument is a criterion for recognizing the trivial knot (i.e. K is trivial if $X(K) \sim S^1$), it applies also when $n = 2$, by Freedman's Unknotting Theorem. (It is easy to see that if K is a 2-knot with finitely generated commutator subgroup π' then K has a finite factorization into irreducibles, for the Grushko-Neumann theorem places an upper bound on the number of nontrivial free factors of π'. If $K = \#K_i$ is a factorization with more terms than this bound then at least one of the summands K_i must have group Z and so be trivial).

For each $n \geqslant 3$ there are n-knots which have several distinct factorizations into irreducibles [BHK 1981]. Essentially nothing is known about uniqueness (or otherwise) of factorization when $n = 2$.

A more general method of combining two knots is the process of forming satellites. Although this construction arose in the classical case [Sch 1953], where it is intimately connected with the notion of torus decomposition, we shall describe only the higher dimensional version of [Ka 1983]. Let K_1 and K_2 be n-knots (with $n > 1$) and let γ be a simple closed curve in $X(K_1)$, with a product neighbourhood U. Then $S^{n+2}-int\,U$ is homeomorphic to $S^n \times D^2$ and so we may find a homeomorphism h which carries $S^{n+2}-int\,U$ onto a product neighbourhood of K_2. The knot $\Sigma(K_2;K_1,\gamma) = hK_1$ is called the *satellite* of K_1 about K_2 relative to γ. We also call K_2 a *companion* of K_1. If either $\gamma = 1$ or K_2 is trivial then $\Sigma(K_2;K_1,\gamma) = K_1$. The group of a satellite knot may be computed by means of van Kampen's Theorem (cf. Chapter 3).

Fibred knots

A *Seifert hypersurface* for K is a locally flat, oriented codimension 1 submanifold of S^{n+2} with (oriented) boundary K. By a standard argument these always exist. As we shall make little use of Seifert hypersurfaces in this book, we shall only outline the argument briefly. (We shall however use the phrase "Seifert *manifold*" below in the sense of closed 3-manifold foliated by circles). Using obstruction theory, it may be shown that the projection $pr_2 j^{-1}:\partial X(K) \to S^n \times S^1 \to S^1$ extends to a map $p:X \to S^1$ [Ke 1965]. By topological transversality, we may assume that the inverse image $p^{-1}(1)$ is a bicollared, proper codimension 1 submanifold of X [Qu 1982]. The union $p^{-1}(1) \cup j(S^n \times [0,1])$ is then a Seifert hypersurface for K. If a 2-knot has a Seifert surface which is a once-punctured connected sum of lens spaces and copies of $S^1 \times S^2$ then it is reflexive [Gl 1962].

In general there is no canonical choice of a Seifert hypersurface. However there is one important special case. An n-knot K is *fibred* if we may find such a map p which is the projection of a fibre bundle (i.e. if every point of S^1 is a regular value of p). The sphere S^{n+2} is then swept out by copies of the Seifert hypersurface obtained from $p^{-1}(1)$, which are disjoint except at their common boundary K. The bundle is determined by the isotopy class of the characteristic map, which is a self homeomorphism θ of the fibre F of p. Indeed $X(K)$ is the mapping torus $F\times_\theta S^1 = F\times[0,1]/\sim$, where $(f,0) \sim (\theta(f),1)$ for all f in F. The isotopy class of θ is called the (geometric) monodromy of the bundle. It is easy to see that such a map p extends to a fibre bundle projection $q:M(K) \to S^1$. The fibre $\hat{F} = F\cup D^{n+1}$ of q is called the *closed fibre* of K, and $M(K)$ is the mapping torus of the closed monodromy. Conversely, if $n \geqslant 2$ and $M(K)$ fibres over S^1 then we may assume that the characteristic map fixes an $(n+2)$-disc pointwise, and we see that K is fibred. (An analogous result is true when $n = 1$ [Ga 1987]). Many of our examples below shall arise as a result of surgery on a simple closed curve in such a mapping torus. (For instance Cappell and Shaneson construct their pairs of distinct n-knots with homeomorphic exteriors by starting with the mapping torus of a self homeomorphism of $(S^1)^{n+1}$).

A 1-knot K is fibred if and only if π' is free [St 1962]. In high dimensions we may apply Farrell's fibration theorem to obtain a criterion for an n-knot with $n \geqslant 4$ to be fibred [Fa 1970]. This applies also when $n = 3$, provided the knot group is in the class of groups for which 4-dimensional surgery and s-cobordism theorems are known (cf. [Fr 1983]). Little is known when $n = 2$. It is conceivable that every 2-knot whose commutator subgroup is finitely generated and torsion free may be fibred. However we shall show in Chapter 8 that if the 3-dimensional Poincaré conjecture is true then there are 2-knots whose commutator subgroup is $Z/3Z$ which are not fibred.

If K_1 and K_2 are fibred then so is their sum, and the closed fibre of $K_1\#K_2$ is the connected sum of the closed fibres of K_1 and K_2. However in the absence of an adequate criterion for a 2-knot to fibre, we do not know whether every summand of a fibred knot is fibred. In view of the unique factorization theorem for oriented 3-manifolds one might hope that there would be a similar theorem for fibred 2-knots. However the fibre of an irreducible 2-knot need not be an irreducible 3-manifold. (For instance a spun trefoil knot is an irreducible fibred 2-knot, but its closed fibre is $S^2\times S^1\#S^2\times S^1$).

No nontrivial 2-knot which is fibred with monodromy of odd order is reflexive [Pl 1986]. (See also [HP 1988]).

Spinning and twist spinning

The first nontrivial examples of higher dimensional knots were given by Artin [Ar 1925]. We may paraphrase his original idea as follows. As the half 3-space $R^3 = \{ <w,x,y,z> \text{ in } R^4 \mid w = 0,\ z \geqslant 0 \}$ is spun about the axis $A = \{ <0,x,y,0> \}$ it sweeps out the whole of R^4, and any arc in R^3 with endpoints on A sweeps out a 2-sphere. This construction has been extended in several ways.

If K is an n-knot, we may choose a small $(n+2)$-disc B^{n+2} which meets K in an n-disc B^n such that (B^{n+2}, B^n) is homeomorphic to the standard pair. Let $(S^{n+2}, K)_o = (S^{n+2} - int\, B^{n+2}, K - int\, B^n)$. Then the (Artin) *spin* of K is the $(n+1)$-knot $\sigma_1 K = \partial((S^{n+2}, K)_o \times D^2)$. The *p-superspin* of K is the $(n+p)$-knot $\sigma_p K = \partial((S^{n+2}, K)_o \times D^{p+1})$. This makes sense for any $p \geqslant 0$. In particular, $\sigma_0 K = K\# -K$. If $p > 0$ then $\pi\sigma_p K = \pi K$. In general superspinning is distinct from iterated spinning (cf. [Ca 1970]). Since the $(p+1)$-disc evidently has an orientation reversing involution, all p-superspun knots are -amphicheiral. The p-superspin of a fibred knot is fibred, and the p-superspin of the sum of two knots is the sum of their p-superspins.

For our purposes, another modification devised by Fox (cf. [Fo 1966]) is of more interest. He incorporated a twist into Artin's original construction. Let r be an integer and choose B^{n+2} meeting K as above. Then $S^{n+2} - int\, B^{n+2} = D^n \times D^2$, and we may choose the homeomorphism so that $\partial(K - int\, B^n)$ lies in $\partial D^n \times \{0\}$. Let ρ_θ be the self homeomorphism of $D^n \times D^2$ that rotates the D^2 factor through θ radians. Then $\cup \rho_{r\theta}(K - int\, B^n) \times \{\theta\}$ is a submanifold of $(S^{n+2} - int\, B^{n+2}) \times S^1$ homeomorphic to $D^n \times S^1$ and which is standard on the boundary. Therefore

$$(S^{n+3}, \tau_r K) = ((S^{n+2} - int\, B^{n+2}) \times S^1, \cup \rho_{r\theta}(K - int\, B^n) \times \{\theta\}) \cup (S^{n+1}, S^{n-1}) \times D^2$$

is an $(n+1)$-knot, called the *r-twist spin* of K.

The 0-twist spin is the Artin spin, i.e. $\tau_0 K = \sigma_1 K$. The group of $\tau_r K$ is obtained from πK by adjoining the relation making the rth power of (any) meridian central. Zeeman discovered the remarkable fact that if $r \neq 0$ then $\tau_r K$ is fibred with geometric monodromy of order dividing r, and the closed fibre is the r-fold branched cyclic cover of S^{n+2}, branched over K [Ze 1965]. Hence $\tau_1 K$ is always trivial. The r-twist spin of the sum of two knots is the sum of their r-twist spins. The twist spin of a −amphicheiral knot is −amphicheiral, while twist spinning interchanges invertibility and +amphicheirality [Li 1985]. The 2-twist spin of any knot is reflexive [Mo 1983, Pl 1984']. (More precisely, if $K = \tau_2 k$ then K^* is isotopic to rK). On the other hand, if $r > 2$ then no non-trivial cyclic branched cover of an r-twist spin of a simple 1-knot is reflexive [HP 1988].

For other formulations and extensions of twist spinning see [GK 1978, Li 1979, Pl 1984', Mo 1983, Mo 1984]. In [Mo 1986] it is shown how to represent twist spins of classical knots by means of hyperplane cross sections (as in [Fo 1962, Lo 1981]).

Slice and ribbon knots

An n-knot K is a *slice* knot if there is an $(n+1)$-knot which meets the equatorial S^{n+2} of S^{n+3} transversally in K; if the $(n+1)$-knot can be chosen to be trivial then K is K is *doubly slice*. As Kervaire showed that all even-dimensional knots are slice [Ke 1965], this notion is of little interest in connection with 2-knots. However not all slice knots are doubly slice, and no adequate criterion is yet known. The 0-spin $\sigma_0 K = K\#-K$ of any knot K is a slice of the 1-twist spin of K and so is doubly slice [Su 1971].

An n-knot K is a *ribbon* knot if it is the boundary of an immersed $(n+1)$-disc Δ in S^{n+2} whose only singularities are transverse double points, the double point set being a disjoint union of discs. All ribbon knots are slice. It remains an open question as to whether every slice 1-knot is ribbon, but in every higher dimension there are slice knots which are not ribbon [Hi 1979]. Given such a "ribbon" $(n+1)$-disc Δ in S^{n+2} the cartesian product $\Delta \times D^p \subset S^{n+2} \times D^p \subset S^{n+2+p}$ determines a ribbon $(n+1+p)$-disc in S^{n+2+p}. All higher dimensional ribbon knots derive from ribbon 1-knots by this process [Ya 1977]. As the p-disc has an orientation reversing involution, this easily implies that all ribbon n-knots with $n \geqslant 2$ are −amphicheiral. The 0-spin of a 1-knot is a ribbon 2-knot. Each ribbon 2-knot has a Seifert hypersurface which is a once-punctured connected sum of copies of $S^1 \times S^2$ and therefore is reflexive [Ya 1969]. (See [Su 1976] for more on such geometric properties of ribbon 2-knots).

An n-knot K is a *homotopy ribbon* knot if it bounds a properly embedded $(n+1)$-disc in D^{n+3} whose exterior W has a handlebody decomposition consisting of 0, 1 and 2-handles. The boundary of W is clearly $M(K)$. The dual decomposition of W relative to its boundary has only $(n+1)$- and $(n+2)$-handles, and so the inclusion of M into W is n-connected. (The definition of "homotopically ribbon" for 1-knots given in [GK: Problem 4.22] requires only that this latter condition be satisfied). Every ribbon knot is homotopy ribbon [Hi 1979]. A nontrivial twist spin of a 1-knot is never homotopy ribbon [Co 1983]. (See also Chapter 3).

Links

Knot theory is the paradigm for the general problem of codimension 2 embeddings of connected manifolds in manifolds. Although we do not intend to stray far from our concentration on 2-knots, we shall occassionally point out where the results described below may be extended to more general situations. In particular, similar questions arise in connection with the groups of links and of homology spheres, so we shall describe these briefly.

A *μ-component n-link* is a locally flat embedding $L:\mu S^n \to S^{n+2}$. It has exterior $X(L) = S^{n+2} - L \times int\, D^2$ and its group is $\pi L = \pi_1(X(L))$. A link L is *trivial* if it bounds a collection of μ disjoint $(n+1)$-discs in S^{n+2}. It is *split* if each of its components lies in an $(n+2)$-disc in S^{n+2} which is disjoint from the other components, and it is a *boundary* link if it bounds a collection of μ disjoint orientable hypersurfaces in S^{n+2}. Clearly a trivial link is split, and a split link is a boundary link; neither implication can be reversed (if $\mu > 1$). Each knot is a boundary link, and many arguments with knots that depend upon Seifert hypersurfaces extend readily to boundary links. The notions of slice and ribbon links are natural extensions of the corresponding notions for knots (cf. [H: Chapter II]).

A μ-component n-link is a boundary link if and only if there is a homomorphism from πL to $F(\mu)$, the free group of rank μ, which carries a set of meridians (one for each component) to a free basis for $F(\mu)$; such a homomorphism can be realized by a continuous map from $X(L)$ to $\vee^{\mu}S^1$, the wedge of μ circles. If $n \neq 2$ the link is trivial if and only if this map is $(n+1)/2$-connected [Gu 1972]. When $n = 2$ the correct criterion for triviality is unknown: it is plausible that every μ-component 2-link whose group is freely generated by meridians is trivial. (The condition on the meridians is necessary [Po 1974]). Also unknown is a good criterion for a 2-link to split, and whether every 2-link is slice. (Even-dimensional boundary links are always slice [Gu 1972]). The exterior of an n-link with $n > 1$ and more than 1 component never fibres over the circle [H: Theorem VIII.4]. Every ribbon n-link with $n > 1$ is a sublink of a ribbon link whose group is free [H: Theorem II.1].

An *homology m-sphere* is a closed m-manifold with the integral homology of S^m. More generally, we may consider μ-component links of homology m-spheres in an homology $(m+2)$-sphere with $\mu \geqslant 0$. Unfortunately while Kervaire's characterization of high dimensional knot groups (Theorem 1 of Chapter 2) extends readily to a characterization of high dimensional link groups, a crucial assumption in one of our principal theorems in Chapter 3 below is that $\chi(X(L)) = 0$, which is only possible when $\mu = 1$, i.e. when we are considering knots (in homology 4-spheres). (We may instead propose the following less standard situations to which our methods probably extend without difficulty. These are when we consider embeddings of one or several tori $S^1 \times S^1$ in $S^3 \times S^1$ or $S^1 \times S^1 \times S^1 \times S^1$, or of pairs of 2-spheres in $S^2 \times S^2$).

For recent work on embeddings of other surfaces in S^4, see [Li 1981, FKV 1988, Li 1988].

Chapter 2 THE KNOT GROUP

Kervaire characterized the groups of n-knots (for each $n \geq 3$) as the finitely presentable groups G with $G/G' \cong Z$, $H_2(G;Z) = 0$ and which are the normal closure of a single element [Ke 1965]. These conditions are also necessary when $n = 1$ or 2, but are then no longer sufficient. The group of a nontrivial 1-knot has geometric dimension 2 and has one end [Pa 1957]. The main concern of this book is with the intermediate case $n = 2$. In this case Kervaire showed also that if a group satisfying the above conditions has deficiency 1 then it is the group of a 2-knot. However not every 2-knot group has deficiency 1. Subsequently several people observed independently and approximately simultaneously that not every high dimensional knot group is a 2-knot group. Their arguments all used duality in the infinite cyclic cover of the exterior of the knot.

In this chapter we shall review these results and various of their extensions and applications. In the next chapter we shall show that by using duality in more general covering spaces we can get much stronger results.

Kervaire's conditions

If S is a subset of a group G we shall let $\langle\langle S\rangle\rangle_G$ denote the *normal closure* of S in G, the smallest normal subgroup of G which contains S. The *weight* of a group is the minimum number of elements in a subset whose normal closure is the whole group. If G has weight 1 and g is an element such that $\langle\langle g\rangle\rangle_G = G$ then we call g a *weight element* for G, and its conjugacy class a *weight class* for G.

As the group π of an n-knot K is the fundamental group of a compact $(n+2)$-manifold it is finitely presentable. By Alexander duality $H_1(X;Z) \cong Z$ and $H_2(X;Z) = 0$. Therefore $\pi/\pi' \cong Z$ (by the Hurewicz theorem) and $H_2(\pi;Z) = 0$ (since it is the cokernel of the Hurewicz homomorphism from $\pi_2(X)$ to $H_2(X;Z)$, by a theorem of Hopf [Ho 1942]). Moreover π has weight 1, for if μ is a meridian, represented by a simple closed curve on ∂X then $X \cup_\mu D^2$ is a deformation retract of $S^{n+2} - \{*\}$ and so is simply connected. (Alternatively we may observe that any loop γ in X bounds a singular disc in S^{n+2} which may be assumed to meet K

transversely in finitely many points; γ is then homotopic in X to a product of conjugates of meridians, bounding transverse discs near these points). Kervaire showed that, conversely, any group satisfying these conditions is the group of some n-knot, for each $n \geqslant 3$. In fact it is sufficient to show that each such group can be realized by a 3-knot, for $\pi\sigma_p K = \pi K$ for all $p > 0$, and so we may call such groups 3-knot groups.

Theorem 1 [Ke 1965] *A group G is a 3-knot group if and only if it is finitely presentable, $G/G' \cong Z$, $H_2(G;Z) = 0$ and G has weight* 1.

Proof The conditions are necessary, by the above remarks. Let P be a finite presentation for G, with g generators and r relators, and let $C(P)$ be the corresponding 2-dimensional cell complex, with one 0-cell, g 1-cells and r 2-cells. For each $n \geqslant 2$ we may embed $C(P)$ in R^{n+3}. Choose such an embedding and let U be a regular neighbourhood of the image. Then ∂U is a closed s-parallelizable $(n+2)$-manifold, and the inclusion of ∂U into U is an n-connected map, since $C(P)$ has codimension $n+1$ in U. Therefore $\pi_1(\partial U) = \pi_1(U) = \pi_1(C(P)) = G$ and $H_k(\partial U;Z) = H_k(C(P);Z)$ for $k \leqslant n$. Since $C(P)$ is a finite 2-dimensional complex $H_2(\partial U;Z) = H_2(C(P);Z)$ is free abelian of finite rank. Since $H_2(G;Z) = 0$ the Hurewicz map from $\pi_2(\partial U)$ to $H_2(\partial U;Z)$ is onto [Ho 1942]. Therefore if $n \geqslant 3$ we can represent a basis of $H_2(\partial U;Z)$ by embedded spheres, which have trivial normal bundles as ∂U is s-parallelizable. On performing surgery on these spheres we obtain a closed orientable $(n+2)$-manifold V with fundamental group G (since the surgered spheres have codimension $n \geqslant 3$), and which has the homology of $S^{n+1} \times S^1$. If we now perform surgery on a weight element of $\pi_1(V) \cong G$, we obtain a simply connected homology $(n+2)$-sphere, which is therefore S^{n+2} by the validity of thc high dimensional Poincaré conjecture, and which contains an n-knot (the cocore of the surgery) with group G. □

The only points at which the high dimensionality was used were where we wished to use surgery to kill $H_2(\partial U;Z)$, while retaining $\pi_1(V) \cong G$, and when we invoked the high dimensional Poincaré conjecture.

The following extension to the case $n = 2$ is due to Kervaire, apart from the appeal to the subsequent work of Freedman. (Recall that if P is a finite presentation of G, with g generators and r relators, then the deficiency of P is $def\ P = g-r$, and $def\ G$ is the maximal deficiency of all finite presentations of G).

Addendum [Ke 1965] *If G satisfies the hypotheses of Theorem 1 and also has a presentation of deficiency 1 then G is a 2-knot group.*

Proof Let β be the rank of $H_2(\partial U;Z)$. Then $\beta = 1-1+\beta = \chi(C(P)) = 1-g+r = 1-def\ P$. Therefore $def\ P \leqslant 1$, and $def\ P = 1$ if and only if $H_2(\partial U;Z) = 0$. In the latter case we do not need to surger any 2-spheres. Since there is no difficulty in surgering 1-spheres in dimension 4, and since the 4-dimensional (TOP) Poincaré conjecture is true [Fr 1982], the above construction gives a knot in S^4. □

It may be shown that any 3-knot group with a presentation of deficiency 1 is in fact the group of a homotopy ribbon 2-knot, by using such a presentation to construct a 5-dimensional handlebody $W = D^5\cup\{h_i^1\}\cup\{h_j^2\}$ with $\pi_1(\partial W) = \pi_1(W) \cong G$ and $\chi(W) = 0$. Adjoining another 2-handle h along a loop representing a weight class for G gives a homotopy 5-ball B with 1-connected boundary. Thus ∂B is a homotopy 4-sphere, and the boundary of the cocore of the 2-handle h is clearly a homotopy ribbon 2-knot with group G.

It is easy to see that if $def\ G = 1$ then $weight\ G = 1$ implies that $G/G' \cong Z$, and that $G/G' \cong Z$ implies that $H_2(G;Z) = 0$. The conditions $G/G' \cong Z$, $H_2(G;Z) = 0$ and $weight\ G = 1$ are otherwise independent. (For instance, $Z*SL(2,5)$ satisfies the first two conditions, but does not have weight 1 [GR 1962]; if G is the (metabelian) group with presentation $\langle a,t \mid t^2at^{-2} = atat^{-1} = tat^{-1}a\rangle$ (i.e. the group of the closed 3-manifold obtained by 0-framed surgery on the trefoil knot) then $G/G' \cong Z$ and G has weight 1, but $H_2(G;Z) \cong Z$, and any finite cyclic group satisfies the last two conditions but not the first).

If G is a group with $G/G' \cong Z$ and deficiency 1 then the first nonzero elementary ideal $E_1(G)$ is principal, and so G'/G'' is torsion free (cf. Chapter IV of [H]). Therefore the group of the 2-twist spin of the trefoil knot does not have deficiency 1, for it has commutator subgroup cyclic of order 3. Thus the deficiency condition is too stringent in general.

Kervaire gave analogous characterizations of the groups of high dimensional links and homology spheres. As the proofs are similar to that of Theorem 1 we shall just state his results.

Theorem A [Ke 1965'] *A group G is the group of a μ-component 3-link if and only if it is finitely presentable,* $G/G' \cong Z^{\mu}$, $H_2(G;Z) = 0$ *and G has weight μ. If moreover G has a presentation of deficiency μ then it is the group of a μ-component 2-link.* □

The group G of a μ-component 1-link has weight μ and $G/G' \cong Z^{\mu}$, but $H_2(G;Z)$ is only 0 (equivalently, $def\ G = \mu$) if the link is completely splittable [H: Theorem I.2].

If we combine Theorem A with Gutiérrez' characterization of high dimensional boundary links [Gu 1972] we find that the groups of such links are distinguished by the additional condition that G maps onto $F(\mu)$, the free group of rank μ, with a set of weight elements for G mapping to a free basis for $F(\mu)$. If we drop the condition on the weight of G in Theorem A we obtain a characterization of the groups of links of μ n-spheres in homology $(n+2)$-spheres (for $\mu \geqslant 0$). In particular we have

Theorem B [Ke 1969] *A group G is the group of an homology m-sphere (for each $m \geqslant 5$) if and only if it is finitely presentable,* $G = G'$ *and* $H_2(G;Z) = 0$*. If moreover G has a presentation of deficiency 0 then it is the group of an homology 4-sphere.* □

Theorem B is easier in so far as there is no need to appeal to the h-cobordism theorem (in order to recognize the standard sphere). Little else is known about the groups of homology 4-spheres. There are many finite perfect groups with presentations of deficiency 0. (Note that a perfect group G with a presentation of deficiency 0 is superperfect, i.e. $H_i(G;Z) = 0$ for $i = 1$ and 2). For instance the groups $SL(2,q)$ for $q = 8$ or q a power of an odd prime ($q \neq 3$) are examples of such groups [CR

1980]. It is not known whether deficiency 0 is a necessary condition. In particular can $I^*\times I^*$ be the group of an homology 4-sphere, where $I^* = SL(2,5)$ is the binary icosahedral group? (Note that I^* is the only finite group that is the group of an homology 3-sphere).

The more general problem of codimension 2 embeddings of n-manifolds V in S^{n+2} does not lead to new groups unless some component of V has nontrivial first homology, since $H_i(S^{n+2}-V;Z)$ is isomorphic to $H_{i-1}(V;Z)$ for $i = 1$ and 2, by Alexander duality in S^{n+2} and Poincaré duality in V.

The commutator subgroup

In our later arguments we shall often first identify the commutator subgroup π' of a knot group π, and so we shall reformulate the Kervaire conditions. We say that an automorphism ϕ of a group G is *meridianal* if $<<g^{-1}\phi(g) \mid g \text{ in } G>>_G = G$. If H is a characteristic subgroup of G then clearly ϕ induces a meridianal automorphism of the quotient G/H. In particular the induced endomorphism $H_1(\phi)-1$ of $H_1(G;Z) = G/G'$ is then onto. If G is solvable, an automorphism satisfying the latter condition is meridianal, for a solvable perfect group is trivial.

The symbol $G\times_\phi Z$ denotes the semidirect product of Z with the normal subgroup G, in which some element t whose image modulo G is the generator 1 of the quotient Z acts on G via $tgt^{-1} = \phi(g)$ for all g in G. If ϕ is meridianal then the semidirect product clearly has weight 1 and commutator subgroup G. In general there may be infinitely many such semidirect products. They may be classified in terms of conjugacy classes in the outer automorphism group $Out(G)$.

Lemma 1 *Let ϕ and θ be meridianal automorphisms of G. Then they determine isomorphic semidirect products if and only if ϕ is conjugate to θ or θ^{-1} in $Out(G)$.*

Proof Suppose that h is an isomorphism from $\pi_\phi = G\times_\phi Z$ to $\pi_\theta = G\times_\theta Z$. Since G is the commutator subgroup in both cases, h induces an isomorphism $e:Z \to Z$ of the quotients, for some $e = \pm 1$. Let t,u be fixed elements of the semidirect products π_ϕ, π_θ respectively which map to 1 in Z. Then $h(t) = u^e g$ for some g in G. Therefore $h(\phi(h^{-1}(j))) = h(th^{-1}(j)t^{-1}) = u^e gjg^{-1}u^{-e} = \theta^e(gjg^{-1})$ for all j in G. Thus ϕ is conjugate to θ^e in $Out(G)$.

Conversely if ϕ is conjugate to θ^e in $Out(G)$ there is an f in $Aut(G)$ and a g in G such that $\phi(j) = f^{-1}\theta^e f(gjg^{-1})$ for all j in G. Then $F(t) = u^e f(g)$, $F|G = f$ defines an isomorphism from π_ϕ to π_θ. □

Theorem 2 [HK 1978, Le 1978] *A finitely presentable group π is a 3-knot group if and only if $\pi = \pi' \times_\theta Z$ for some meridianal automorphism θ of π' such that $H_2(\theta)-1$ is an automorphism of $H_2(\pi';Z)$.*

Proof Let π be the group of an n-knot K and let t be a meridian for π. Since $\pi = \langle\langle t \rangle\rangle_\pi$, every element of π is a product of powers of terms gtg^{-1}, where we may assume that g is in π'. Therefore every element of π' is a product of commutators $[gt^{\pm 1}g^{-1}, ht^{\pm 1}h^{-1}]$ with g,h in π'. It is easily verified that all such commutators are in $\langle\langle k^{-1}tkt^{-1} \mid k \text{ in } \pi' \rangle\rangle_{\pi'}$, which is therefore all of π'. Then

$$\langle\langle g^{-1}\theta(g) \mid g \text{ in } \pi' \rangle\rangle_{\pi'} = \langle\langle g^{-1}tgt^{-1} \mid g \text{ in } \pi' \rangle\rangle_{\pi'} = \pi',$$

and so θ is meridianal. It is clear that $\pi = \pi' \times_\theta Z$.

As observed in Chapter 1, it follows from the Wang sequence for the projection of X' onto X that multiplication by $t-1$ acts invertibly on $H_i(X';Z) = H_i(X;\Lambda)$ for $i > 0$. Now $H_2(\pi';Z)$ is the cokernel of the Hurewicz map from $\pi_2(X')$ to $H_2(X';Z)$, which commutes with the action of the covering group, and so multiplication by $t-1$ gives an automorphism (namely $H_2(\theta)-1$) of $H_2(\pi';Z)$ also.

Conversely any such group π has weight 1, abelianization Z, and on applying the Wang sequence for the projection of $K(\pi',1)$ onto $K(\pi,1)$ we find that $H_2(\pi;Z) = 0$, and so π is a 3-knot group. □

Hausmann and Kervaire have expressed the condition that $\pi' \times_\theta Z$ be finitely presentable in terms of π' having a "finite Z-dynamic presentation" [HK 1978].

When π' is an abelian group A the condition that the automorphism be meridianal reduces to the action of $t-1$ on $A = H_1(A;Z)$ being invertible. Moreover in that case the group $H_2(A;Z)$ may be identified with the exterior product $A\wedge A$ [R: page 334]. Levine and Weber have made explicit the conditions on the pair (A,t) in the following way [LW 1978]. If A is a finitely generated Λ-module let $\Delta_0(A)$ be the highest common factor of the ideal $E_0(A)$ in Λ, and for each prime p let $\Delta(A\otimes F_p)$ be the generator of the ideal $E_0(A\otimes F_p)$ in $\Lambda\otimes F_p = F_p[t,t^{-1}]$. (See [H: Chapter III] for these elementary ideals). We shall say that a Laurent polynomial $\lambda(t)$ in $F_p[t,t^{-1}]$ is *asymmetric* if $\lambda(1) \neq 0$ and the highest common factor of $\lambda(t)$ and $\lambda(t^{-1})$ is 1 or $t+1$.

Theorem C [LW 1978] *A finitely presentable group π of the form $A\times_\phi Z$ is a 3-knot group if and only if $\Delta(A\otimes F_p)$ is asymmetric for each prime p.* □

Such a semidirect product is always finitely generated as a group, if A is finitely generated as a Λ-module. Trotter has shown that $A\times_\phi Z$ is finitely presentable if and only if either the highest or the lowest coefficient of $\Delta_0(A)$ is ± 1 [Tr 1974]. (In fact he assumes also that A is Z-torsion free, but as the Z-torsion of a finitely generated Λ-module on which $t-1$ acts invertibly is finite (see Lemma 2 below), this assumption is unnecessary. See also [BS 1978]).

The condition for a finitely generated abelian group A to be a 3-knot commutator subgroup (i.e. to admit such an automorphism t) has been given explicitly by Hausmann and Kervaire.

Theorem D [HK 1978] *A finitely generated abelian group A is the commutator subgroup of a 3-knot group if and only if*

(i) $dim_Q A\otimes Q \neq 1$ *or* 2;

(ii) $r(2^k) \neq 1$ *or* 2, *for all* k ; *and*

(iii) $r(3^k) = 1$ *for at most one value of* k.

Here $r(p^k)$ is the number of summands isomorphic to Z/p^kZ in the *primary decomposition of the torsion subgroup of A.* □

In general a metabelian knot group is an ascending HNN extension over a finitely generated abelian base [BS 1978]. Yoshikawa has considered arbitrary HNN extensions with finitely generated abelian base [Yo 1986]. (Geometrically this corresponds to the knot having a "minimal" Seifert hypersurface with abelian fundamental group).

Theorem E [Yo 1986] *Let H be a finitely generated abelian group and* $\phi: I \rightarrow J$ *an isomorphism between two subgroups of H. Then the HNN extension* $G = H*_\phi$ *(with presentation* $\langle H,t \mid tit^{-1} = \phi(i)$ *for all i in* $I\rangle$*) is a 3-knot group if and only if*

(*i*) $\Delta((G'/G'')\otimes F_p)$ *is asymmetric for each prime p ; and*

(*ii*) *rank I = rank J = rank H and the orders of the quotients* $H/(I+zH)$ *and* $H/(J+zH)$ *are relatively prime, where zH is the torsion subgroup of H.*

Moreover a finitely generated abelian group H is the base of some 3-knot group (i.e., admits some such pair of subgroups) if and only if it satisfies conditions (ii) and (iii) of Theorem D. □

We shall not give proofs for these results, as we present them only for contrast with the lower dimensional cases. The commutator subgroup of a nontrivial 1-knot group has a nonabelian free subgroup (the image of the fundamental group of an incompressible Seifert surface for the knot) and so is never abelian. In fact it is either free of finite rank or the direct limit of iterated free products of free groups amalgamated over free subgroups, and then is not finitely generated [N: Theorem 4.5.1]. We shall see that if the commutator subgroup of a 2-knot group is abelian then it is either Z^3 or $Z[½]$, the dyadic rationals, or is cyclic of odd order.

Yoshikawa has also shown that any 2-knot group which is an HNN extension with finitely generated abelian base is either metabelian or has base of rank 1. For each integer ν, the group with presentation $\langle a,t \mid ta^\nu t^{-1} = a^{\nu+1}\rangle$ is an example of the latter kind. (Note that this is metabelian only if $\nu = -2, -1, 0$ or 1). It is not known whether there are any other such 2-knot groups. It is however clear from Yoshikawa's work that no nontrivial classical knot group is an HNN extension with finitely generated abelian base. (Since classical knot groups have cohomological dimension 2 and symmetric Alexander polynomials, this also follows directly from Theorem E together with [B: Corollary 6.7]).

Duality in the infinite cyclic cover

The simplest example of a 3-knot group which is not a 2-knot group is due to Farber [Fa 1977]. He showed that the Z-torsion subgroup of the first homology of the infinite cyclic cover of the exterior of a 2-knot supports a nondegenerate symmetric bilinear pairing into Q/Z, for which the covering group acts isometrically. (This was also discovered by Levine [Le 1977]). We shall construct this pairing using duality in the infinite cyclic cover (following Levine's argument), but we shall refer to the original papers of Farber and Levine for full details and for proofs that the pairing is symmetric. (We do not need to know this). Let zA denote the Z-torsion subgroup of a Λ-module A.

Lemma 2 [Ke 1965] *Let A be a finitely generated Λ-module on which $t-1$ acts invertibly. Then A is a Λ-torsion module, and the Z-torsion subgroup zA is a finite Λ-submodule.*

Proof If we tensor A with the field of fractions $Q(t)$ of Λ we get 0, and as A is finitely generated it must be a torsion module. It is clear that zA is a submodule, and that $t-1$ acts invertibly on zA. Moreover since Λ is noetherian zA is also finitely generated, and so has finite exponent m as an abelian group. Suppose first that m is prime. Then zA is a finitely generated module over the P.I.D. $F_m[t,t^{-1}]$, on which $t-1$ acts invertibly, and so must be finite. In general we may induct on the number of primes dividing m, for if $p.zA$ and $zA/p.zA$ are both finite then so is zA. □

If N is a Λ-module we shall let $e^qN = \overline{Ext^q(N,\Lambda)}$, for all $q \geqslant 0$. If N is finitely generated then so is e^qN (for all q), for N has a finite free resolution, since Λ is noetherian of global dimension 2 and $K_0(\Lambda) = 0$. If N is a Λ-torsion module then $e^0N = 0$ and $e^1N \cong$

$\overline{Hom_\Lambda(N,Q(t)/\Lambda)}$. (This follows on applying the functor $\overline{Hom_\Lambda(N,-)}$ to the short exact sequence $0 \to \Lambda \to Q(t) \to Q(t)/\Lambda \to 0$, in which the middle term is an injective Λ-module).

Lemma 3 [Le 1977] *(i) Let A be a finitely generated Λ-module such that $zA = 0$. Then $e^2A = 0$.*

(ii) Let B be a finitely generated Λ-torsion module. Then e^1B has no nontrivial finite submodule.

Proof (i) By the exactness of localization $e^2A\otimes Q = \overline{Ext_{Q\Lambda}(A\otimes Q,Q\Lambda)}$, which is 0 since $Q\Lambda = Q[t,t^{-1}]$ is a P.I.D. Therefore since e^2A is finitely generated there is an integer $m > 0$ with $m.e^2A = 0$. Since $zA = 0$ the sequence $0 \to A \xrightarrow{m} A \to A/m.A \to 0$ is exact, and so we obtain an exact sequence $e^2A \to e^2A \to e^3(A/m.A) = 0$. Therefore $e^2A = 0$.

(ii) This is proven in [H: Theorem III.11]. □

Corollary *Let A be a finitely generated Λ-module on which $t-1$ acts invertibly. Then $e^2A = e^2zA$ and is finite.*

Proof By Lemma 3, $e^2(A/zA) = 0$. The first assertion now follows on applying the functor $\overline{Hom_\Lambda(-,\Lambda)}$ to the short exact sequence $0 \to zA \to A \to A/zA \to 0$. Since e^2zA is a finitely generated Z-torsion Λ-module on which $t-1$ acts invertibly, it is finite by Lemma 2. □

We have assumed that $t-1$ acts invertibly, as it simplifies our argument, but the corollary remains true without this assumption.

Theorem 3 [Fa 1977, Le 1977] *Let K be a 2-knot with group π and let $A = \pi'/\pi'' = H_1(M(K);\Lambda)$. Then $H_2(M(K);\Lambda) \cong e^1A$, and there is a nondegenerate Z-bilinear pairing $[\ ,\ ]:zA\times zA \to Q/Z$ for which t acts as an isometry: $[t\alpha,t\beta] = [\alpha,\beta]$ for all α and β in zA.*

proof By Poincaré duality in the infinite cyclic cover of $M(K)$ and the Universal Coefficient spectral sequence there are isomorphisms $A = \overline{H^3(M;\Lambda)}$

and $H_2(M;\Lambda) = \overline{H^2(M;\Lambda)} \cong e^1H_1(M;\Lambda) = e^1A$ and a short exact sequence $0 \to e^2H_1(M;\Lambda) \to \overline{H^3(M;\Lambda)} \to e^1H_2(M;\Lambda) \to 0$. Thus there is a monomorphism $\delta: e^2A \to A$ with cokernel $e^1H_2(M;\Lambda)$. Since A and $H_2(M;\Lambda)$ are finitely generated Λ-torsion modules on which $t-1$ acts invertibly, it follows from Lemma 2 and its corollary that δ gives an isomorphism $\eta: zA \to e^2zA$. Now Levine shows that if B is any finite Λ-module there is an isomorphism $\xi_B: e^2B \cong Hom_Z(B,Q/Z)$ where the latter group has the Λ-module structure determined by $t.f(b) = f(t^{-1}b)$ for all b in B and $f: B \to Q/Z$. We define $[\alpha,\beta] = \xi_{zA}(\eta(\alpha))(\beta)$ for all α, β in zA. The pairing is clearly Z-bilinear, and is nondegenerate since η and ξ_{zA} are isomorphisms. Finally $[t\alpha,t\beta] = \xi_{zA}(t\eta(\alpha))(t\beta) = (t\xi_{zA}(\eta(\alpha))(t\beta) = \xi_{zA}(\eta(\alpha))(\beta) = [\alpha,\beta]$. □

When K is fibred, with closed fibre $\hat{F}$, this pairing is just the standard linking pairing on the torsion subgroup of $H_1(\hat{F};Z)$, together with the automorphism induced by the monodromy. In Chapter 7 we shall construct an isomorphism analogous to ξ_B, in a more general setting.

Corollary [Le 1978] *$H_2(\pi';Z)$ is a quotient of $\overline{Hom_\Lambda(\pi'/\pi'',Q(t)/\Lambda)}$.*

Proof This is an immediate consequence of the theorem, since $H_2(\pi';Z)$ is a quotient of $H_2(M';Z) = H_2(M;\Lambda)$ by Hopf's theorem. □

If π has deficiency 1 then $H_2(\pi';Z) = 0$ (cf. [B: Section 8.5] or [H: page 42]). Levine observed that if a 3-knot group π has deficiency d then π'/π'' has deficiency $\geqslant d-1$ as a Λ-module and that therefore $e^2(\pi'/\pi'')$ can be generated by at most $1-d$ elements. It then follows from Theorem 3 that the group of the sum of m copies of the 2-twist spin of the trefoil knot has deficiency $1-m$ [Le 1978]. There are in fact irreducible 2-knots whose groups have deficiency $1-m$, for each $m \geqslant 0$ [Ka 1983].

Farber gave the following example. Let π be the group with presentation $\langle a,t \mid tat^{-1} = a^2, a^5 = 1\rangle$. Then π' is cyclic of order 5, $t-1$ acts as the identity and $H_2(\pi';Z) = 0$, so π is a 3-knot group. However it is easy to see that if $l(\ ,\ )$ is any nondegenerate Z-bilinear pairing on π'

with values in Q/Z then $l(t\alpha,t\beta) = 4l(\alpha,\beta) = -l(\alpha,\beta)$ for all α,β in π'. Thus t cannot be an isometry and so π is not a 2-knot group.

Centres

Since the commutator subgroup of a classical knot group π can contain no nontrivial abelian normal subgroup [N: Chapters IV,V], any nontrivial abelian normal subgroup of π must map injectively to $\pi/\pi' = Z$ and so be central. Burde and Zieschang have shown that in fact the knot must then be a torus knot [BZ 1966].

In high dimensions the situation is quite different. Hausmann and Kervaire have shown that every finitely generated abelian group A is the centre of some n-knot group, for each $n \geqslant 3$ [HK 1978']. They observe that if P is a finitely presentable superperfect group and G is a knot group then by the Künneth theorem $H_i(G\times P;Z) = H_i(G;Z)$ for $i \leqslant 2$, and $G\times P$ has centre $\zeta(G\times P) = \zeta G\times\zeta P$. If moreover there is an element p in P such that the subgroup $[p,P]$ generated by $\{[p,x] = pxp^{-1}x^{-1} \mid x \text{ in } P\}$ is the whole of P, and if g is a weight element for G, then gp is a weight element for $G\times P$, which is therefore a 3-knot group. More generally if $P_1, \cdots P_r$ are r such superperfect groups (with such elements p_i in P_i) then $G\times P_1\times\cdots\times P_r$ is a 3-knot group with centre $\zeta G\oplus\zeta P_1\oplus\cdots\oplus\zeta P_r$. Thus to obtain the result of Hausmann and Kervaire it shall suffice to take G to be a knot group with trivial centre (for instance the group of the figure eight knot) and to show that each cyclic group is the centre of some such group P. For the infinite cyclic group we may take P to be the fundamental group of a suitable Brieskorn homology 3-sphere, for instance the group with presentation $\langle a,b \mid a^7 = b^3 = (ab)^2\rangle$, with p the element represented by a. For the cyclic group of order 2 we may use $I^* = SL(2,5)$ (which is also the group of a Brieskorn homology 3-sphere), with p any noncentral element. If $k \geqslant 3$ and F is any finite field containing k distinct k^{th} roots of unity (other than F_4) then $SL(k,F)$ is a finite superperfect group (cf. [M: pages 78,94]) with centre cyclic of order k; we may take p to be the elementary matrix e_{12}^1 (with entry 1 in the (1,2) position). Hausmann and Kervaire gave more general constructions and showed that each such A is the centre of infinitely many 3-knot groups.

They left open the questions as to whether the centre need be finitely generated, and what the centre of a 2-knot group may be. In Chapters 3 and 4 we shall show that the centre of a 2-knot group is either torsion free of rank 2 or has rank at most 1. (Examples are known with centre Z^2, $Z\oplus(Z/2Z)$, Z or $Z/2Z$). Whether the centre of a 2-knot group must be finitely generated is related to the analogous question for 3-dimensional Poincaré duality groups, which seems difficult.

If $F(\mu)$ is a free group of rank $\mu \geqslant 2$, and if $P_1, \cdots P_r$ are finitely presentable superperfect groups with elements p_i as above, then the group $F(\mu)\times P_1\times \cdots \times P_r$ is the group of a μ-component boundary n-link (for each $n \geqslant 3$) with centre $\oplus\zeta P_i$. Thus the centre of a μ-component n-link can be any finitely generated abelian group, if $\mu \geqslant 1$ and $n \geqslant 3$. We shall show in Chapter 3 that the centre of a μ-component 2-link with $\mu > 1$ must be a torsion group.

Minimizing the Euler characteristic

We saw above that a 3-knot group π must satisfy certain homological conditions, notably $H_2(\pi;Z) = 0$, and that if the stronger, combinatorial condition $def\,\pi = 1$ also holds then π is a 2-knot group. Analogous situations arise in attempting to characterize the groups of 2-links, homology 4-spheres etc. Now every finitely presentable group is the fundamental group of some closed orientable 4-manifold. Thus we may define a new invariant which may help bridge the gap between necessary homological and sufficient combinatorial conditions by

$$q(G) = \min\{\chi(M) \mid M \textit{ a closed orientable 4-manifold with } \pi_1(M) = G\}.$$

Hausmann and Weinberger observed that this invariant is well behaved with respect to subgroups of finite index, since the Euler characteristic is multiplicative in finite coverings, and it is this property that makes $q(\ )$ useful.

For any space M and field F let $\beta_i(M;F)$ be the ith Betti number of M with coefficients F, and for any group G let $\beta_i(G;F) = \beta_i(K(G,1);F)$. Then we have the following estimates.

Theorem 4 [HW 1985] *Let G be a finitely presentable group. Then* $2(1-\beta_1(G;F))+\beta_2(G;F) \leqslant q(G) \leqslant 2(1-def\ G)$.

Proof Let M be any space with fundamental group G. Then $H_1(M;F) = H_1(G;F)$ and $H_2(M;F)$ maps onto $H_2(G;F)$. If M is an orientable 4-manifold then by Poincaré duality $\chi(M) = 2(1-\beta_1(M;F))+\beta_2(M;F)$. These two observations give the first inequality.

If P is a finite presentation for G and $C(P)$ is the corresponding 2-complex then we may embed $C(P)$ in R^5. The boundary V of a regular neighbourhood is then a closed s-parallelizable 4-manifold with $\pi_1(V) = G$ and $\chi(V) = 2(1-def\ G)$, and so we get the second inequality. □

Corollary *Let H be a subgroup which has finite index in G. Then $2(1-\beta_1(H;F))+\beta_2(H;F) \leqslant [G:H]q(G)$.*

Proof If M is a closed 4-manifold with $\pi_1(M) = G$ and M_H is the cover with group H then $\chi(M_H) = [G:H]\chi(M)$. Hence $q(H) \leqslant [G:H]q(G)$. □

We may now refine the Addendum to Theorem 1.

Theorem 5 *Let π be a 3-knot group. Then $q(\pi) \geqslant 0$, and is 0 if and only if π is a 2-knot group.*

Proof Since $\beta_1(\pi;F) = 1$ and $\beta_2(\pi;F) = 0$ for any field F, the first assertion is clear. If $q(\pi) = 0$ then there is a closed 4-manifold M with $\pi_1(M) = \pi$ and $\chi(M) = 0$. Surgery on a weight class then gives a simply connected closed 4-manifold Σ with $\chi(\Sigma) = 2$, which must therefore be S^4, and the cocore of the surgery is a 2-knot with group π. The coverse is clear, for if K is any 2-knot then $\chi(M(K)) = 0$. □

Corollary *Let π be a 2-knot group such that π' is finite. Then every abelian subgroup of π' is cyclic, and so π has periodic cohomology.*

Proof Suppose that π' has a noncyclic abelian subgroup. Then there is a prime p and an abelian subgroup A isomorphic to $(Z/pZ)^2$, and π has a subgroup H of finite index which is isomorphic to $A\oplus Z$. Since $\beta_1(H;F_p) = 3$ and $\beta_2(H;F_p) = 5$ we have a contradiction to the corollary of Theorem 4. The final assertion follows from [CE: page 262]. □

The above proof for this corollary is from [HW 1985]. In fact the finite group π' must have cohomological period 4, as we shall show in Chapter 4. Note that for each $n \geqslant 0$ the group $Z\times(I^*)^n$ is a 3-knot group whose commutator subgroup has centre $(Z/2Z)^n$, and so it can only be a 2-knot group if $n = 0$ or 1.

Hausmann and Weinberger also used the invariant $q(\ )$ to give the first examples of homology 5-sphere groups which are not the groups of homology 4-spheres. (They gave two infinite families of examples, one consisting of finite groups and the other of torsion free groups).

Deficiency and cohomological dimension

It is well known that if K is a classical knot then $\widetilde{X}(K)$ is contractible [Pa 1957]. (Equivalently, by duality in the universal cover, either $\pi K \cong Z$ or it has one end). As X is a 3-manifold with nonempty boundary, it collapses to a finite 2-complex [N: Chapter III]. In particular, π has cohomological dimension at most 2. Moreover, π has a Wirtinger presentation of deficiency 1, that is, a presentation of the form $<x_i\ ,\ 0\leqslant i\leqslant g \mid x_j = w_j x_0 w_j^{-1},\ 1\leqslant j\leqslant g>$, in which each relation asserts the conjugacy of two of the generators. In this section we shall examine some of the connections between these properties.

Since the Artin spin of a 1-knot is a ribbon 2-knot every 1-knot group is the group of some ribbon 2-knot. By an elementary handle sliding argument it may be seen that any ribbon n-link (with $n \geqslant 2$) is a sublink of a ribbon link whose group is free [H: Theorem II.1]. It follows that the group of a μ-component ribbon n-link (with $n \geqslant 2$) has a (Wirtinger) presentation of deficiency μ, which is optimal, since the group has weight μ. (This was first proven by Yajima for the groups of ribbon 2-knots [Ya 1969]). Conversely any group of weight μ with a (Wirtinger) presentation of deficiency μ is the group of a μ-component sublink of a (ribbon) n-link with group free, for each $n \geqslant 2$ [H: Theorem II.3]. Since the group of a homotopy ribbon n-knot (with $n \geqslant 2$) is the group of a $(n+3)$-manifold W which can be built with 0-, 1- and 2-handles only and which has Euler characteristic 0, such groups also have deficiency 1.

Levine showed that if π has a presentation P such that the presentation of the trivial group obtained by adjoining the relation killing a meridian to P is AC-equivalent to the empty presentation then π is the group of a smooth knot in the standard 4-sphere. According to Yoshikawa, π has such a presentation if and only if it has a Wirtinger presentation of deficiency 1 [Yo 1982']. (See also [Si 1980] for connections between Wirtinger presentations and the condition that $H_2(\pi;Z) = 0$).

A group has *(finite) geometric dimension* 2 if it is the fundamental group of a (finite) aspherical 2-complex, but is not free. Every such group has cohomological dimension 2. It is an open question as to whether every (finitely presentable) group of cohomological dimension 2 has (finite) geometric dimension 2 (cf. [W': Problem D.4]). The following partial answer to this question was first obtained by Beckmann under the further assumption that G has type FL (cf [Dy 1987']).

Theorem 6 *Let G be a finitely presentable group. Then G has finite geometric dimension 2 if and only if it has cohomological dimension 2 and deficiency $\beta_1(G;Q)-\beta_2(G;Q)$.*

Proof Let P be a presentation for G with g generators and r relators, and let $C(P)$ be the corresponding 2-complex. Then $def\ P = 1-\chi(C(P)) = \beta_1(C(P);Q)-\beta_2(C(P);Q) \leqslant \beta_1(G;Q)-\beta_2(G;Q)$, and the necessity of the conditions is clear. The cellular chain complex of the universal cover $\widetilde{C}(P)$ may be viewed as a finite chain complex of free left $Z[G]$-modules, and so there is an exact sequence

$$0 \to L = \pi_2(C(P)) \to Z[G]^r \to Z[G]^g \to Z[G] \to Z \to 0.$$

As $c.d.G = 2$, the image of $Z[G]^r$ in $Z[G]^g$ is projective, by Schanuel's lemma. Therefore the inclusion of L into $Z[G]^r$ splits, and L is projective. Since $c.d.G = 2$, the Hattori-Stallings rank of L is concentrated on the conjugacy class <1> of the identity [Ec 1986], and so the Kaplansky rank of L is the dimension of $Q\otimes_{Z[G]}L$. If $def\ P = \beta_1(G;Q)-\beta_2(G;Q)$ then $Q\otimes_{Z[G]}L = 0$ and so $L = 0$, by a theorem of Kaplansky. (See Section 2 of [Dy 1987] for more details on the properties and interrelations of the various notions of rank). Hence $C(P)$ is aspherical. □

Suppose now that G has weight μ and a presentation P of deficiency μ, and let $C(P)$ be the corresponding 2-complex. The 2-complex obtained by adjoining μ 2-cells to $C(P)$ along loops representing a μ-element subset of G whose normal closure is the whole group is simply connected and has Euler characteristic 1, and so is contractible. Therefore if the Whitehead conjecture is true the subcomplex $C(P)$ must also be aspherical, and so G is free or has finite geometric dimension 2. (In particular, this is so if G is a 1-relator group (cf. [Ly 1950, Go 1981]), or is locally indicable [Ho 1982] or if it has no nontrivial superperfect normal subgroup [Dy 1987]). Thus Theorem 6 and the Whitehead conjecture together imply that a 3-knot group has finite geometric dimension 2 if and only if it has deficiency 1, in which case it is a 2-knot group.

If the commutator subgroup of a 2-knot group π with deficiency 1 is finitely generated must it be free? This is so if π is a classical knot group [N: Theorem 4.5.1] or if $c.d.\pi = 2$ and π' is almost finitely presentable [B: Corollary 8.6].

Sphericity of the exterior

The oustanding property of the exterior of a classical knot is that it is aspherical. In contrast, the exterior of a higher dimensional knot is aspherical only when it has the homotopy type of a circle, in which case the knot must be trivial [DV 1973]. The proof that we shall give is due to Eckmann, who also showed that the exterior of a higher dimensional link with more than one component is never aspherical [Ec 1976]. (The exterior of a 1-link is aspherical if and only if the link is unsplittable).

Theorem 7 [DV 1973, Ec 1976] *Let K be an n-knot, for some $n > 1$, such that $X(K)$ is aspherical. Then K is trivial.*

Proof Let $i:\partial X \to X$ be the natural inclusion. Since $n > 1$ we have $\pi_1(\partial X) = Z$ and since X is aspherical i factors through $S^1 = K(\pi_1(\partial X),1)$, i.e. there are maps $j:\partial X \to S^1$ and $h:S^1 \to X$ with i homotopic to hj. Therefore the map i^* from $H^{n+1}(X;Z[\pi])$ to $H^{n+1}(\partial X;i^*Z[\pi])$ factors through $H^{n+1}(S^1;h^*Z[\pi])$ and so is the 0 map. By Poincaré duality, $H^{n+1}(X,\partial X;Z[\pi]) = H_1(X;Z[\pi]) = H_1(X;Z) = 0$, and so $H^{n+1}(X;Z[\pi]) = 0$.

By Poincaré duality again, $H_1(X,\partial X;Z[\pi]) = 0$ and so $H_0(\partial X;i^*Z[\pi])$ maps isomorphically to $H_0(X;Z[\pi])$. But this means that the induced cover of ∂X is connected and so $i_*:\pi_1(\partial X) \to \pi_1(X)$ is onto. Therefore $\pi = Z$ and so $X \sim S^1$, since it is aspherical. The theorem now follows from the unknotting criterion. □

Both Dyer and Vasquez and Eckmann prove somewhat more general results. Eckmann also observes that the full strength of the assumption of asphericity is not needed for the above theorem. Together [Sw 1976] and [Du 1985] imply that if $i_n:\pi_n(\partial X) \to \pi_n(X)$ is the 0 map then K must be trivial.

In the next chapter we shall show that the closed manifold $M(K)$ obtained by surgery on a 2-knot is often aspherical. The group of such a knot has one end; however Gonzalez-Acuña and Montesinos have given examples of 2-knot groups with infinitely many ends, of which the simplest has presentation $<a,b,t \mid a^7 = b^3 = 1,\ bab^{-1} = a^2,\ tbt^{-1} = b^2>$ [GM 1978]. (This group is evidently an HNN extension of the metacyclic group generated by $\{a,b\}$; it may also be viewed as the free product of an isomorphic metacyclic group with the group of the 2-twist spun trefoil knot, amalgamated over a subgroup of order 3).

Weight elements, classes and orbits

Two knots K and K_1 have homeomorphic exteriors if and only if there is a homeomorphism from $M(K_1)$ to $M(K)$ which carries the conjugacy class of a meridian of K_1 to that of K (up to inversion). In fact if M is any closed orientable 4-manifold with $\chi(M) = 0$ and with $\pi = \pi_1(M)$ of weight 1 then surgery on a weight class gives a 2-knot with group π. Moreover, if t and u are two weight elements and f is a self homeomorphism of M such that u is conjugate to $f_*(t^{\pm 1})$ then surgeries on t and u lead to knots whose exteriors are homeomorphic (via the restriction of a self homeomorphism of M isotopic to f). Thus the natural invariant to distinguish between knots with isomorphic groups is not the weight class, but rather the *weight orbit*: the orbit of a weight element under the automorphisms of the group.

A refinement of this notion is useful in distinguishing between oriented knots. If w is a weight element for π then we shall call the set $\{\alpha(w) \mid \alpha \text{ in } Aut(\pi),\ \alpha(w) \equiv w \bmod \pi'\}$ a *strict* weight orbit for π. A strict weight orbit determines a transverse orientation for the corresponding knot (and its Gluck reconstruction). An orientation for the ambient sphere is determined by an orientation for $M(K)$. If K is invertible or +amphicheiral then there is an orientation preserving (respectively, orientation reversing) self homeomorphism of M which reverses the transverse orientation of the knot, i.e. carries the strict weight orbit to its inverse. Similarly, if K is −amphicheiral there is an orientation reversing self homeomorphism of M which preserves the strict weight orbit.

Theorem 8 *Let G be a group of weight 1 with $G/G' \cong Z$, and let t be an element of G whose image generates G/G'. For each g in G let c_g be the automorphism of G' induced by conjugation by g. Then*

(i) t is a weight element if and only if c_t is meridianal;

(ii) two weight elements t, u are in the same weight class if and only if there is an element g of G' such that $c_u = c_g c_t c_g^{-1}$;

(iii) two weight elements t, u are in the same strict weight orbit if and only if there is an automorphism d of G' such that $c_u = dc_t d^{-1}$ and $dc_t d^{-1} c_t^{-1}$ is an inner automorphism;

(iv) if t and u are weight elements then u is conjugate to $(g''t)^{\pm 1}$ for some g'' in G''.

Proof The verification of (i–iii) is routine. If t and u are weight elements then, up to inversion, u must equal $g't$ for some g' in G'. Since $t-1$ acts invertibly on G'/G'' we have $g' = khth^{-1}t^{-1}$ for some h in G' and k in G''. Let $g'' = h^{-1}kh$. Then $u = g't = hg''th^{-1}$. □

An immediate consequence of this theorem is that if t and u are in the same strict weight orbit then c_t and c_u have the same order, and their images in $Out(G')$ are equal. Moreover if C is the centralizer of c_t in $Aut(G')$ then the strict weight orbit of t contains at most $\#(Aut(G')/C.Inn(G')) \leqslant \#Out(G')$ weight classes. In general there may be infinitely many weight orbits [Pl 1983']. However, if π is metabelian the weight class (and hence the weight orbit) is unique up to inversion, by part (iv) of the theorem.

Fibred 2-knots

Many of the most interesting examples arise from 4-manifolds which fibre over the circle; the knots are then fibred knots. The counter example of Farber given above shows that some restriction on torsion is needed for the next result.

Theorem 9 *Let π be a torsion free 3-knot group such that π' is the fundamental group of a closed orientable 3-manifold N whose factors are Haken, hyperbolic or Seifert fibred. Then π is the group of a fibred 2-knot with closed fibre N.*

Proof Let $N = P\#R$ where P is a connected sum of copies of $S^1\times S^2$ and R has no such factors. The meridianal automorphism of $\pi' \cong \pi_1(N)$ may be realized by a self homotopy equivalence g of N [Sw 1974]. By the Splitting Theorem of [La 1974], g is homotopic to a connected sum of homotopy equivalences between the irreducible factors of R with a self homotopy equivalence of P. By our assumptions, these maps are homotopic to homeomorphisms. (Cf. [Wa 1968, Mo 1968, Sc 1983] and [La 1974] respectively). Thus we may assume that g is a self homeomorphism of N. Surgery on a weight class in the mapping torus of g gives a fibred 2-knot with closed fibre N and group π. □

Since N is determined by π' and homotopy implies isotopy if N is a connected sum of Haken manifolds and copies of $S^1\times S^2$ [La 1974], or is hyperbolic [Mo 1968] or Seifert fibred [Sc 1985, BO 1986], the mapping torus of g is unique up to homeomorphism among fibred 4-manifolds with such a fibre.

The r-twist spin of a 1-knot K is a fibred 2-knot, and the factors of its closed fibre are Haken, hyperbolic or Seifert fibred. With some exceptions for small values of r, the factors are aspherical, and $S^1\times S^2$ is never a factor. (See Chapter 5 for a more precise statement).

If the group π of a fibred 2-knot has deficiency 1 then $H_2(\pi';Z) = 0$ and so the closed fibre is a connected sum of copies of $S^1\times S^2$ with a homology 3-sphere. The homology 3-sphere is simply connected if and only if π' is free. Cochran has shown that if the closed fibre is $\#^r(S^1\times S^2)$ the knot is homotopy ribbon, and that conversely the closed fibre of a fibred homotopy ribbon 2-knot is a connected sum of copies of $S^1\times S^2$ with a homotopy 3-sphere [Co 1983]. For fibred ribbon knots this follows from an argument of Trace. If V is any Seifert hypersurface for an n-knot K then the embedding of V in X extends to an embedding of $\hat{V} = V\cup D^{n+1}$ in M, which lifts to an embedding in M'. Since the image of $\hat{V}$ in $H_{n+1}(M;Z)$ is Poincaré dual to a generator of $H^1(M;Z) = Hom(\pi,Z) = [M,S^1]$, its image in $H_{n+1}(M';Z) \cong Z$ is a generator. If K is fibred M' is homotopy equivalent to the closed fibre $\hat{F}$, so there is a degree 1 map from $\hat{V}$ to $\hat{F}$ and hence to any factor of $\hat{F}$. If moreover $n = 2$ and $\pi_1(V)$ is free then each prime factor of $\hat{F}$ must be $S^1\times S^2$ or a homotopy 3-sphere, by Lemma 1 of [Tr 1986]. In particular, since ribbon 2-knots have such Seifert hypersurfaces no nontrivial fibred ribbon 2-knot is a twist spin.

Ruberman has used a similar idea together with the fact that the Gromov norm of a 3-manifold does not increase under degree 1 maps to show that no fibred 2-knot with closed fibre a hyperbolic 3-manifold can have a Seifert hypersurface V such that $\hat{V}$ is a graph manifold [Ru 1987]. (In particular, $\hat{V}$ cannot be a Seifert fibred 3-manifold or $\#^r(S^1\times S^2)$). He also observes that the "Seifert volume" of Brooks and Goldman [BG 1984] may be used in a similar way to show that some fibred 2-knots whose closed fibre has an $\widetilde{SL}(2,R)$-geometric structure cannot be ribbon knots.

Chapter 3 LOCALIZATION AND ASPHERICITY

Although the exterior $X(K)$ of a nontrivial 2-knot K is never aspherical, in many cases the closed 4-manifold $M(K)$ *is* aspherical. This is so for instance if K is one of the Cappell-Shaneson knots, or (with finitely many exceptions) if it is the q-twist spin of a prime classical knot for some $q > 2$. In this chapter we shall show that this is true whenever the knot group has a nontrivial torsion free abelian normal subgroup and is cohomologically 1-connected at infinity. (We shall also consider how the latter condition might fail). The knot group is then a Poincaré duality group of formal dimension 4 and orientable type, or PD_4^+-group for short. (We shall also show that, conversely, if πK is a PD_4-group and $H^1(\pi';Z/2Z) \neq 0$ then it is orientable and $M(K)$ is aspherical).

Our argument is based on the idea of embedding the group ring $Z[\pi]$ in a larger ring R in which an annihilator for the augmentation module becomes invertible and for which nontrivial stably free modules have well defined strictly positive rank, with R^n having rank n. Using Poincaré duality and the cohomological condition on π we then find that the equivariant homology with coefficients R of a closed orientable 4-manifold with group π is concentrated in degree 2 and is stably free as an R-module. Its rank may then be computed by an Euler characteristic counting argument. When π has a nontrivial torsion free abelian normal subgroup the existence of such an overring is guaranteed by a remarkable lemma of Rosset [Ro 1984].

This strategy works in considerably greater generality, provided we forgo some information about torsion. For instance, although solvable groups have abelian normal subgroups, there may be finitely presentable infinite solvable groups in which no such subgroup is torsion free. In order to get around this problem we may factor out the maximal locally-finite normal subgroup. (This idea is due to Kropholler). The quotient of a 2-knot group by such a subgroup is then usually a PD_4^+-group over Q.

Rosset's Lemma

The keystone of the argument of this chapter (and hence of the whole book) is the following lemma of Rosset.

Lemma [Ro 1984] *Let G be a group with a torsion free abelian normal subgroup A, and let S be the multiplicative system $Z[A]-\{0\}$ in $Z[G]$. Then the (noncentral!) localization $R = S^{-1}Z[G]$ exists and has the property that each nontrivial finitely generated stably free R-module has well defined strictly positive rank, with R^n having rank n. Moreover R contains $Z[G]$ as a subring, and is flat as a $Z[G]$-module, and if A is nontrivial then $R \otimes_{Z[G]} Z = 0$.* □

The prototype of such a result was given by Kaplansky who showed that for any group G the group ring $Z[G]$ has this strong "invariant basis number" property (cf. [K: page 122]). Rosset observes that the multiplicative system S satisfies the Ore conditions (cf. [P: page 146]) and so the localization exists and is flat; if a is a nontrivial element of A then $a-1$ is in S and annihilates the augmentation module Z. Beyond this his argument follows that of Kaplansky in making use of properties of C^*-algebras. (In [Hi 1981] we stated such a lemma for the case when A is central, and tried to derive it algebraically from Kaplansky's Lemma, but there was a gap in our argument).

On the evidence of his work on 1-relator groups, Murasugi conjectured that the centre of a finitely presentable group other than Z^2 of deficiency $\geqslant 1$ is infinite cyclic or trivial, and is trivial if the group has deficiency > 1, and he showed that this is true for the groups of 1-links [Mu 1965]. (The classical knots and links whose groups have nontrivial centre have been determined by Burde, Zieschang and Murasugi [BZ 1966, BM 1970]). As a corollary to our first application of Rosset's Lemma we shall show that a stronger conjecture is true for a large class of finitely presentable groups, including all those with a central element of infinite order.

Theorem 1 *Let W be a finite connected 2-dimensional cell complex such that $G = \pi_1(W)$ has a nontrivial torsion free abelian normal subgroup A. Then $\chi(W) \geqslant 0$, and $\chi(W) = 0$ if and only if W is aspherical.*

Proof Let $\widetilde{W}$ be the universal cover of W with the equivariant cell structure. Then the cellular chain complex of $\widetilde{W}$ may be viewed as a finite chain complex of free left $Z[G]$-modules $C_* = 0 \to C_2 \to C_1 \to C_0 \to 0$ where C_i has rank c_i, the number of i-cells of W. Since $\widetilde{W}$ is simply connected, $H_0(C_*) = Z$ and $H_1(C_*) = 0$, while $H_2(C_*) = H_2(\widetilde{W};Z) = \pi_2(W)$ is a submodule of C_2.

Let S be the multiplicative system $Z[A]-\{0\}$ in $Z[G]$. Since A is nontrivial $Z_S = 0$, and so on localizing the chain complex C_* we obtain an exact sequence $0 \to H_2(C_*)_S \to C_{2S} \to C_{1S} \to C_{0S} \to 0$ from which it follows that $H_2(C_*)_S$ is a stably free $Z[G]_S$-module of rank $c_0-c_1+c_2 = \chi(W)$, which must therefore be nonnegative. As $H_2(C_*)$ is a submodule of C_2, which embeds in C_{2S}, it is 0 if and only if its localization is 0. Thus $\widetilde{W}$ is contractible if and only if $\chi(W) = 0$. □

The assumption that W be 2-dimensional is not needed in order to show that W aspherical implies that $\chi(W) = 0$; this is in fact Rosset's application of his lemma. Gottlieb obtained the first such result in the case of an aspherical complex whose fundamental group had nontrivial centre [Go 1965]. The lemmas of Kaplansky and Rosset have been used in related ways in connection with the Whitehead conjecture on the asphericity of subcomplexes of 2-dimensional $K(G,1)$-complexes (cf. [Dy 1987] and the references therein).

Corollary 1 *If a finitely presentable group G has a nontrivial torsion free abelian normal subgroup then it has deficiency at most* 1. *If G has deficiency* 1 *and is not Z then it has finite geometric dimension* 2. *If moreover the centre ζG of G is nontrivial and G is nonabelian then ζG is infinite cyclic and the commutator subgroup G' is free.*

Proof Let P be a finite presentation for G and let $C(P)$ be the related 2-complex. Then $\chi(C(P)) = 1-def\ P$ and so the theorem implies directly all but the last two assertions, which then follow from [B: Theorem 8.8]. □

As the groups of classical links are all torsion free and have deficiency $\geqslant 1$, this corollary implies immediately the above-mentioned results of Murasugi.

Corollary 2 *If G has deficiency 1 and is nonabelian and $G/G' \cong Z$, then G' is infinite.*

Proof Let C be a finite 2-complex with $\pi_1(C) = G$ and $\chi(C) = 0$. If G' were finite then G would have an infinite cyclic subgroup of finite index. The corresponding covering space of C would be a finite 2-complex with fundamental group Z and Euler characteristic 0. It is easy to see that the universal cover of such a complex must be contractible, and so G must be torsion free, and therefore infinite cyclic. □

This corollary does not really need Rosset's Lemma, for it is sufficient to work with the commutative ring Λ. Note that this gives a quick proof that if the group π of a nontrivial classical knot K has finitely generated commutator subgroup then it has one end. It then follows easily from Poincaré duality that $X(K)$ is aspherical.

A cyclic branched cover of S^3, branched over a knot K, is the connected sum of the cyclic branched covers of the prime factors of K. These are irreducible, and cannot be $S^1 \times S^2$ [Pl 1984]. Therefore the commutator subgroup of $\pi\tau_r K$ is a free product of finite groups and PD_3^+-groups, and is never a nontrivial free group. (Thus if $\pi\tau_r K$ is torsion free and not Z it has cohomological dimension 4). Since $\pi\tau_r K$ has a central element of infinite order, Corollary 1 implies that it cannot have deficiency 1, and so in particular $\tau_r K$ cannot be a nontrivial homotopy ribbon 2-knot (cf. [Co 1983]).

Some of the arguments of the next few chapters may be seen in microcosm in the next theorem. We shall let Φ denote the group with presentation $\langle a,t \mid tat^{-1} = a^2 \rangle$. The *centralizer* of a normal subgroup A of a group G is the kernel of the homomorphism from G to $Aut(A)$ determined by the conjugation action of G on A, and shall be denoted by $C_G(A)$; if A is abelian then it is a central subgroup of its centralizer.

Theorem 2 *Let G be a finitely generated group of cohomological dimension 2 with a nontrivial abelian normal subgroup A. Then either A is infinite cyclic or G is solvable. If $G/G' \cong Z$ then either $G \cong \Phi$ or A*

is central and G' is free; in either case G has deficiency 1.

Proof Suppose that A is not infinite cyclic. Then $c.d.A = c.d.C_G(A) = c.d.G = 2$, so $C_G(A)$ is also abelian, by [B: Theorem 8.8]. If A has rank 1 then $Aut(A)$ is abelian, so $G' \leqslant C_G(A)$ and G is metabelian. Otherwise $A = C_G(A) \cong Z^2$. As $C_G(A)$ with an element of G infinite order modulo $C_G(A)$ would generate a subgroup of cohomological dimension 3, which is impossible, $G/C_G(A)$ must be a torsion group. Since it is finitely generated and is isomorphic to a subgroup of $Aut(A) \cong GL(2,Z)$, it must be finite [K: page 105]. Since G is torsion free it must be Z^2 or the Klein bottle group. In general the solvable groups of cohomological dimension 2 have been determined by Gildenhuys [Gi 1979]. Among them only Φ has abelianization Z.

If $A \cong Z$ then $[G:C_G(A)] \leqslant 2$, so $c.d.C_G(A) = 2$. Therefore $C_G(A)'$ is free, by [B: Theorem 8.8]. If $G/G' \cong Z$ then this free subgroup must be nonabelian for otherwise G would be solvable. Hence $A \cap C_G(A)' = 1$. The subgroup $C_G(A)'$ is normal in G and A maps injectively to $H = G/C_G(A)'$. As H has an abelian normal subgroup of index $\leqslant 2$ and $H/H' \cong Z$, we must in fact have $H \cong Z$. It now follows easily that A is central in G, i.e. that $C_G(A) = G$, and so G' is free. The final observation follows readily. □

Groups with cohomological dimension 2, nontrivial centre and abelianization Z are iterated free products of (one or more) torus knot groups, amalgamated over central subgroups [St 1976].

Acyclic covering spaces

Our second application of Rosset's Lemma leads ultimately to more substantial results on 2-knot groups.

Theorem 3 *Let M be a closed connected orientable 4-manifold with fundamental group G. Suppose that there are normal subgroups $T < U$ of G and a subring R of Q such that $Hom(T/T',R) = 0$, U/T is a non-trivial torsion free abelian group and $H^s(G/T;R[G/T]) = 0$ for $s \leqslant 2$.*

Then $\chi(M) \geq 0$ and the covering space M_T of M with group T is R-acyclic (i.e., $H_i(M;R[G/T]) = 0$ for $i > 0$) if and only if $\chi(M) = 0$.

Proof Let C_* be the cellular chain complex of M_T with coefficients R, with respect to an equivariant cell structure. Then $H_0(C_*) = R$ and $H_1(C_*) = 0$, while since U/T is nontrivial G/T is infinite and M_T is an open 4-manifold, and so $H_4(C_*) = 0$ also. Since $Ext^s_{R[G/T]}(R,R[G/T]) = H^s(G/T;R[G/T]) = 0$ for $s \leq 2$, it follows from the Universal Coefficient spectral sequence that $H^1(G/T,R[G/T]) = 0$ and that $H^2(G/T,R[G/T])$ injects into $Hom_{R[G/T]}(H_2(C_*),R[G/T])$. Therefore by Poincaré duality $H = H_2(C_*)$ injects into $H^* = \overline{Hom_{R[G/T]}(H,R[G/T])}$ and $H_3(C_*) = 0$, and so to prove that M_T is R-acyclic it shall suffice to show that $H^* = 0$.

Let S be the multiplicative system $R[U/T]-\{0\}$ in $R[G/T]$, and let $\Gamma = R[G/T]_S$. Since U/T is nontrivial $R_S = 0$ and so the only nonzero homology of C_{*S} is H_S in degree 2. If B is any left Γ-module then $H^3(Hom_\Gamma(C_{*S},B)) = H_1(B\otimes C_{*S}) = 0$ by Poincaré duality and the Künneth theorem. Hence as in [W: Lemma 2.3] H_S is stably free and we may split the boundary maps of the complex C_{*S} to obtain an isomorphism $H_S \oplus C_{1S} \oplus C_{2S} = C_{0S} \oplus C_{2S} \oplus C_{4S}$, so H_S has rank $\chi(M)$ as a stably free Γ-module. In particular, $\chi(M) \geq 0$.

By Rosset's Lemma, if $\chi(M) = 0$ then $H_S = 0$. Since $R[G/T]$ embeds in Γ, H^* embeds in $H^*_S = \overline{Hom_\Gamma(H_S,\Gamma)}$ and so is also 0. Therefore M_T is R-acyclic. Conversely, if M_T is acyclic then $H = 0$ so $H_S = 0$ and hence $\chi(M) = 0$. □

The orientability of M is not essential for this theorem. The assumption on T is satisfied if T is perfect ($T = T'$) and $R = Z$ or if T is a torsion group and $R = Q$. (In particular it holds when $T = 1$). The condition on the subgroup U could be replaced by the apparently more general condition that $R[G/T]$ have a flat extension with the strong invariant basis number property [C], where some annihilator of the augmentation module R becomes invertible (cf. [Hi 1980]). But we know of no other construction of such extensions of group rings. Note also that if a group has an abelian normal subgroup which is either central or finitely generated then it has a torsion free abelian normal subgroup of the same rank.

A group is a $PD_n^{(+)}$-group over R if it is a Poincaré duality group of formal dimension n over R (and orientable type, i.e. $H^n(G;R[G])$ is isomorphic to the augmentation $R[G]$-module R) [B: Chapter III]. We shall omit the qualification "over R" when $R = Z$.

Corollary 1 *If $\chi(M) = 0$ then the quotient group G/T is a PD_4^+-group over R. Hence rank $U/T \leqslant c.d.U/T \leqslant 4$.*

Proof The chain complex C_* is then a finite free acyclic resolution of the augmentation module R, and by equivariant Poincaré duality in M_T the cohomology $H^i(C_*;R[G/T])$ is R if $i = 4$ and 0 otherwise. Therefore G/T is a PD_4^+-group over R by [B: Proposition 9.2] and the orientability of M. The second assertion then follows from [B: Proposition 4.9]. □

Corollary 2 *If L is a μ-component 2-link with $\mu > 1$ then πL does not have any such pair of subgroups T, U.*

Proof The group πL is the fundamental group of the closed 4-manifold obtained by surgery on L, which has Euler characteristic $2-2\mu$. □

Corollary 3 *If the fundamental group of a closed 4-manifold N has an abelian subgroup A of finite index then $\chi(N) \geqslant 0$, and if $\chi(N) = 0$ then A has rank 1, 2 or 4.*

Proof By passing to a subgroup of finite index we may assume that N is orientable, and that $\pi_1(N) = A$ and is free abelian, of rank β say. We may clearly assume that $\beta > 1$. If $\beta > 2$ the theorem implies that either $\chi(N) > 0$ or N is aspherical, and then $\beta = 4$. If $\beta = 2$ then the localized spectral sequence still collapses, and Poincaré duality implies that the only nonzero localized homology module has rank $\chi(N)$, which therefore must be nonnegative. □

The manifolds $S^1{\times}S^3$, $S^1{\times}S^1{\times}S^2$ and $S^1{\times}S^1{\times}S^1{\times}S^1$ have Euler characteristic 0 and fundamental group free abelian of rank 1, 2 and 4 respectively. (See also Corollary 3 of Theorem 3 of Chapter 7).

Corollary 4 *If π is a virtually abelian 2-knot group then either π' is finite (i.e., π is virtually Z) or π is torsion free and virtually Z^4.*

Proof It is readily verified that a group with abelianization Z is virtually Z if and only if its commutator subgroup is finite, while no such group can be virtually Z^2 or Z^3. □

We shall determine completely such 2-knot groups in Chapters 4 and 6.

The cohomological conditions

In this section we shall show that the cohomological hypotheses of Theorem 3 are automatically satisfied if the group U/T is large enough. If J is a finitely generated group then $e(J)$ (= 0, 1, 2 or ∞) shall denote the number of ends of J. If J is infinite $H^0(J;Z[J]) = 0$ while if J has one end then $H^1(J;Z[J]) = 0$ also [Sp 1949].

Lemma 1 *Let A be an abelian group of rank $r \leqslant \infty$ and $F = Z[A]^{(I)}$ a free A-module. Then $H^i(A;F) = 0$ if $i < r$. If A is torsion free and finitely generated then $H^r(A;F) \cong Z^{(I)}$. If $r < \infty$ but A is not finitely generated then $H^r(A;F) = 0$.*

Proof Let B be a free abelian subgroup of A of finite rank $s \leqslant r$. If $r < \infty$ we may assume B has rank r and if A is finitely generated and torsion free we may take $B = A$. Since B is an FP group and F is free as a B-module, $H^i(B;F) = H^i(B;Z[B])\otimes_{Z[B]}F$ for all i [B: Proposition 2.4]. Therefore $H^i(B;F) = 0$ if $i < s$ and $H^s(B;F) \cong Z\otimes_{Z[B]}F \cong Z[C]^{(I)}$, where $C = A/B$. If A is not finitely generated then C is infinite and so $H^0(C;Z[C]^{(I)}) = 0$ [B: Lemma 8.1]. The lemma now follows on applying the LHS spectral sequence $H^p(C;H^q(B;F)) \Rightarrow H^{p+q}(A;F)$. □

Theorem 4 *Let J be a finitely generated group with $J/J' \cong Z$ and which has an abelian normal subgroup A of rank at least 2. Then $H^s(J;Z[J]) = 0$ for $s \leqslant 2$.*

Proof By Lemma 1 the E_2^{pq} terms of the LHS spectral sequence $H^p(J/A;H^q(A;Z[J])) \Rightarrow H^{p+q}(J;Z[J])$ vanish for $q \leqslant 2$, if A has rank greater than 2, or if it has rank 2 and is not finitely generated. If A is finitely generated and of rank 2 then we may assume that it is free, and $E_2^{pq} = 0$ for $q \leqslant 1$. Moreover J must be infinite (cf. Corollary 4 of Theorem 3), so $E_2^{02} = H^0(J/A;Z[J/A]) = 0$ also. In all cases we may conclude that $H^s(J;Z[J]) = 0$ for $s \leqslant 2$. □

Corollary *If a 2-knot group π has a nontrivial torsion free abelian normal subgroup A then the rank of A is at most 4.*

Proof This is an immediate consequence of Theorems 3 and 4. □

In Chapter 2 we saw that every finitely generated abelian group is the centre of some 3-knot group.

The rank 1 case is somewhat more delicate. In the next theorem we shall let $F(a,t)''$ denote the set of all words in the second commutator subgroup of the free group on $\{a,t\}$. As in Chapter 2, we shall let zA denote the Z-torsion subgroup of an abelian group A.

Theorem 5 *Let J be a finitely generated group with $J/J' \cong Z$ and which has an abelian normal subgroup A of rank 1. If $e(J/A) = 0$ then J' is finite. If $e(J/A) = 1$ and moreover $A \cong Z$ or J is finitely presentable then $H^s(J;Z[J]) = 0$ for $s \leqslant 2$. If $e(J/A) = 2$ then J/zA has a finite normal subgroup N such that the quotient has a presentation $\langle a,t \mid ta^nt^{-1} = a^{n+1}, F(a,t)''\rangle$ for some $n \neq -1$ or 0; if moreover J is finitely presentable we may assume that $zA = 0$ and then $J/N \cong \Phi$.*

Proof If $e(J/A) = 0$ then J/A is finite and so A is finitely generated. Since A is not contained in J' and has rank 1, $A \cap J'$ is finite. Therefore J' is also finite, as $[J':A \cap J'] \leqslant [J:A]$. If $e(J/A) = 1$ and $A \cong Z$ then the

LHS spectral sequence argument of Theorem 4 works here also. If $e(J/A) = 1$ and J is finitely presentable (but A is not necessarily finitely generated), then J is 1-connected at ∞ by Theorem 1 of [Mi 1986] and so $H^s(J;Z[J]) = 0$ for $s \leqslant 2$ by [GM 1986]. Thus the only case we need consider at length is when J/A has 2 ends.

Since the abelianization of J/A is cyclic, it cannot have $Z/2Z*Z/2Z$ as a quotient. Therefore if it has 2 ends it must have a finite normal subgroup N_o with infinite cyclic quotient. Clearly $N_o = J'/A$. Since A/zA is torsion free abelian of rank 1 its group of automorphisms is abelian; since A/zA is normal in J/zA it must be central in J'/zA, and it has finite index there. Therefore the commutator subgroup of J'/zA is finite [R: Theorem 10.1.4]. Hence the set N of elements of J' whose image in J'/zA has finite order is a characteristic subgroup of J whose image in J/zA is finite. Let H be the quotient group. Then H is finitely generated, $H/H' \cong Z$ and H' is torsion free abelian of rank 1.

Let t represent a generator for H/H'. The group ring $\Lambda = Z[H/H']$ acts on H' via conjugation (i.e. $(hH').g = hgh^{-1}$ for g in H' and h in H), and H' is finitely generated as a Λ-module. Since H' is Z-torsion free $Ann(H')$ is principal [H: page 35], and since H' has rank 1 we must have $H' \cong \Lambda/(mt-n)$ for some m, n in Z. Thus (after relabeling if necessary) H has a presentation $<a,t \mid ta^m t^{-1} = a^n, F(a,t)''>$. Since $H/H' \cong Z$ we must have $m-n = \pm 1$.

Now J has a subgroup of finite index which maps onto Z with abelian kernel A. Therefore if J is finitely presentable, this subgroup is a constructible solvable group by [BB 1976] and [BS 1978], and so J is virtually torsion free. We may then assume that $zA = 0$. Moreover H is then also finitely presentable, and so it has an equivalent presentation of the form

$$<a,t \mid ta^m t^{-1} = a^{m+1}, [t^k a t^{-k}, a] = 1, 0 \leqslant k \leqslant r>$$

for some sufficiently large r. By a Reidemeister-Schreier rewriting process [MKS: page 158] we get a presentation

$$<a_n, n \text{ in } Z \mid a_{n+1}^m = a_n^{m+1}, [a_{n+k}, a_n] = 1, 0 \leqslant k \leqslant r, n \text{ in } Z>$$

for H'. Clearly H' may be obtained from the group with presentation

$$<a_0,...a_{r+1} \mid a_{n+1}^m = a_n^{m+1},\ 0\leq n\leq r,\ [a_u,a_v] = [a_v,a_{r+1}] = 1,\ 0\leq u<v\leq r>$$

as a direct limit of a sequence of amalgamated free products, and so contains this group as a subgroup. But this group maps onto the group $Z/mZ*Z/(m+1)Z$ with presentation $<b,c \mid b^m = c^{m+1} = 1>$ via the map sending a_0 to c , a_{r+1} to b and $a_1, \cdots a_r$ to 1, and so can only be abelian if m or $m+1$ is 1. In both cases we then have $J/N = H \cong \Phi$. □

When J is finitely presentable, the last case of Theorem 5 may also be extracted from [Tr 1974] or [HK 1978]. Is the finite presentability of J needed in the case when $e(J/A) = 1$?

Locally-finite subgroups

A group is said to be *locally-finite* if every finitely generated subgroup is finite. In any group the union of all the locally-finite normal subgroups is the unique *maximal* locally-finite normal subgroup [R: Chapter 12.1]. (We use the hyphen to avoid confusion with the stricter notion of locally-(*finite* and *normal*) subgroup). Clearly there are no nontrivial maps from such a group to a torsion free group such as Q.

This notion is of particular interest in connection with solvable groups. Since every finitely generated torsion solvable group is finite [R: 5.4.11], if G is a finitely generated infinite solvable group and T is its maximal locally-finite normal subgroup then G/T is nontrivial. Therefore it has a nontrivial abelian normal subgroup, which is necessarily torsion free. Thus we may apply the theorems of the preceeding two sections to 4-manifolds with such groups. We shall consider solvable 2-knot groups in Chapter 6. For the present we shall give a more general result.

Theorem 6 *Let π be the group of a 2-knot K and T be its maximal locally-finite normal subgroup, and suppose π has a normal subgroup U such that U/T is a nontrivial abelian group. If U/T has rank 1 assume also that $e(\pi/U) < \infty$; if moreover $e(\pi/U) = 1$ assume furthermore that $H^s(\pi/T;Z[\pi/T]) = 0$ for $s \leq 2$. Then either π' is finite or $\pi/T \cong \Phi$ or π/T is a PD_4^+-group over Q.*

Proof We note first that U is torsion free, by the maximality of T. If $H^s(\pi/T;Z[\pi/T]) = 0$ for $s \leqslant 2$ then we may apply Theorem 3 immediately, and Corollary 1 then implies that π/T is a PD_4^+-group over Q. Therefore we may assume that U/T has rank 1 and that $e(\pi/U) = 0$ or 2. If $e(\pi/U) = 0$ then π'/T is finite by Theorem 5, and so $T = \pi'$. Suppose that π' is infinite. Then π cannot have 0 or 2 ends, and it cannot have infinitely many ends since it does not contain a nonabelian free subgroup. Therefore it has 1 end and so $H_3(\widetilde{M};Q) = H_4(\widetilde{M};Q) = 0$. Since π' is locally-finite, $h.d._Q\pi' = 0$ by [B: 4.12]. Therefore the Cartan-Leray spectral sequence relating the homology of $\widetilde{M}$ to that of M' collapses, to give $H_3(M';Q) = 0$. But this contradicts Milnor duality, which implies that $H_3(M';Q) \cong Q$. Therefore π' must be finite.

If $e(\pi/U) = 2$ then by Theorem 5 we may assume that π/T has a presentation of the form $\langle a,t \mid ta^nt^{-1} = a^{n+1}, F(a,t)''\rangle$ for some integer $n \neq -1$ or 0. Although we do not know *a priori* that π/T is finitely presentable, we shall show that the argument of the latter part of Theorem 5 can be adapted to reach the same conclusion. We may assume that π has a presentation $\langle a,t,x_1, \cdots x_p \mid ta^nt^{-1} = a^{n+1}w, r_1, \cdots r_q\rangle$ with the images of the generators x_i and the word w lying in T. Let $G = \pi/\langle\langle x_1, \cdots x_p,w\rangle\rangle$. Then $G/G'' = \pi/T$ and G has a presentation $\langle a,t \mid ta^nt^{-1} = a^{n+1}, r_1, \cdots r_q,w\rangle$, so the relations $\{r_1, \cdots r_q,w\}$ may be assumed to be words in $F(a,t)''$. On adjoining finitely many more relations, we find that unless $n = 1$ there is a subquotient of G' (and hence of π) which is free, contrary to π being locally-finite by metabelian. □

We shall see in Chapter 4 that if π' is infinite then T is either trivial or infinite. We know of no examples in which T is infinite.

PD-groups, asphericity and orientability

If M is an aspherical closed orientable 4-manifold then $\pi_1(M)$ is a PD_4^+-group. Since we may increase the homology of M without changing the fundamental group by taking the connected sum with a 1-connected 4-manifold, the converse is in general false. However it is conceivable that every PD_4^+-group G is the fundamental group of some such aspherical 4-manifold, N say. We then have $\chi(N) = q(G)$. In this section we shall see that in certain circumstances a 4-manifold whose group is a PD_4^+-group must be aspherical.

If $f:M \to N$ is an $(n-1)$-connected degree 1 map between closed orientable $2n$-manifolds with fundamental group G, the only obstruction to its being a homotopy equivalence is $H_n(f) = ker(f_*:\pi_n(M) \to \pi_n(N))$. Arguing as in Theorem 3 we may show that $H_n(f)$ is a stably free $Z[G]$-module of rank $(-1)^n(\chi(M)-\chi(N))$, and so f is a homotopy equivalence if it is an integral homology equivalence. We shall next adapt this argument to a case in which it is not known *a priori* that the map has degree 1.

Theorem 7 *Let N be a closed orientable 4-manifold and $G = \pi_1(N)$. The classifying map $f:N \to K(G,1)$ is a homotopy equivalence if and only if G is a PD_4^+-group and f induces a rational homology equivalence.*

Proof As these conditions are clearly necessary, we need only show that they are sufficient. Let C_*, D_* and E_* be the equivariant cellular chain complexes of the universal covers $\widetilde{N}$, $\widetilde{K}(G,1)$ and $(\widetilde{K}(G,1),\widetilde{N})$ respectively. Since $H^s(G;Z[G]) = 0$ for $s < 4$, Poincaré duality together with the universal coefficient spectral sequence give an isomorphism of $H_2(C_*)$ with $Hom_{Z[G]}(H_2(C_*),Z[G])$ as in Theorem 3, while $H_i(C_*) = 0$ if $s \neq 0$ or 2. Since $Z[G]$ maps monomorphically to $Q[G]$, $H_2(C_*)$ embeds in $Q\otimes H_2(C_*) = H_2(\widetilde{N};Q)$. As $\widetilde{K}(G,1)$ is contractible, the only possible non-trivial homology module of $Q\otimes E_*$ is $Q\otimes H_3(E_*) = H_2(\widetilde{N};Q)$ which is a stably free $Q[G]$-module by [W: Lemma 2.3]. Since f induces a rational homology equivalence the Euler characteristics of N and $K(G,1)$ are equal. As these are also the Euler characteristics of C_* and D_*, and as the sequence $0 \to C_* \to D_* \to E_* \to 0$ is exact, the Euler characteristic of E_* is 0. Therefore the stably free $Q[G]$-module $H_2(\widetilde{N};Q) = Q\otimes H_3(E_*)$ has rank 0 and so must in fact be 0, by Kaplansky's Lemma. Thus $H_2(C_*) = 0$ and f is a homotopy equivalence. □

Corollary *Suppose that G is a PD_4^+-group over Q and that the cohomology ring $H^*(N;Q)$ is generated by $H^1(N;Q)$. Then N is aspherical.*

Proof The classifying map from N to $K(G,1)$ is clearly a rational (co)homology equivalence and so the theorem applies. □

Theorem 8 *Let K be a 2-knot whose group π is a PD_4^+-group over a field F such that $H^1(\pi';F) \neq 0$. Then the classifying map f from $M(K)$ to $K(\pi,1)$ induces a homology equivalence with coefficients F.*

Proof The infinite cyclic covers M' and $K(\pi,1)' = K(\pi',1)$ each satisfy Milnor duality of formal dimension 3 with coefficients F and the lift $f':M' \to K(\pi',1)$ of f classifies $\pi' = \pi_1(M')$. Therefore the induced maps $f'_1:H_1(M';F) \to H_1(\pi';F)$ and $f'^1:H^1(\pi';F) \to H^1(M';F)$ are also isomorphisms. Since $\pi_2K(\pi',1) = 0$ the map f' is 2-connected and so Whitehead's theorem [Sp: page 399] implies that $f'_2:H_2(M';F) \to H_2(\pi';F)$ is an epimorphism. Therefore the map $f'_3[M']\cap:H^1(\pi';F) \to H_2(\pi';F)$ is an epimorphism, since by the projection formula $f'_3[M']\cap c = f'_2([M']\cap f'^1(c))$ for all c in $H^1(\pi';F)$ [Sp: page 254]. By assumption, $H^1(\pi';F) \neq 0$, so by duality $H_2(\pi';F) \neq 0$ and hence $f'_3[M'] \neq 0$. On considering the map induced between the Wang sequences of the projections of M' onto M and $K(\pi',1)$ onto $K(\pi,1)$ we see that $f_4[M] \neq 0$ and so f induces isomorphisms in (co)homology with coefficients F. □

Corollary *Let K be a 2-knot whose group π is a PD_4^+-group such that $\pi' \neq \pi''$. Then $M(K)$ is aspherical.*

Proof Since $\pi' \neq \pi''$ there is a field F such that $H^1(\pi';F) \neq 0$. By Theorem 8 the classifying map has nonzero degree and therefore induces a rational homology equivalence, so the corollary follows from Theorem 7. □

As a contrast we have the following theorem.

Theorem 9 *Let K be a 2-knot whose group* π *has deficiency* 1, *and suppose that* $\pi' \neq \pi''$. *Then* $\pi_2(M(K)) \neq 0$, *and* π *is not a* PD_4^+*-group.*

Proof Since $\pi' \neq \pi''$ there is a field F such that $H^1(M';F) \cong H^1(\pi';F)$ is nonzero. Milnor duality then implies that $H_2(M';F) \neq 0$. On the other hand $H_2(\pi';F) = 0$ since $def\,\pi = 1$ (cf. [B: Section 8.5] or [H: page 42]), and so the Hurewicz map from $\pi_2(M) = \pi_2(M')$ to $H_2(M';F)$ is onto [Ho 1942]. This proves the first assertion, and the second then follows from the Corollary immediately above. □

For 2-knot groups the assumption of orientability is usually redundant.

Theorem 10 *Let K be a 2-knot whose group* π *is a* PD_4*-group such that* $H^1(\pi';Z/2Z) \neq 0$. *Then* $w_1(\pi) = 0$, *i.e.,* π *is of orientable type.*

Proof The classifying map $f:M(K) \to K(\pi,1)$ is a $Z/2Z$-cohomology equivalence by Theorem 8, since every PD_4-group is orientable over $Z/2Z$. The orientation character w_1 of a 4-dimensional Poincaré duality complex is characterized by the Wu formula $w_1 \cup x = \beta_2(x)$ for all $Z/2Z$-cohomology classes x of degree 3 [Sp: page 350]. Therefore $w_1(M) = f^1(w_1(\pi))$. Since M is orientable and f^1 is injective we see that $w_1(\pi) = 0$. □

We may use this theorem to give another example of a 3-knot group which is not a 2-knot group. Let A be a 3×3 integral matrix with $detA = -1$, $\det(A-I) = \pm 1$ and $\det(A^{(2)}-I) = \pm 1$. (Here $A^{(2)}$ is the induced automorphism of $H_2(Z^3;Z) = Z^3 \wedge Z^3$: in classical terms it is the second compound of A. When the first two conditions hold, the third is equivalent to $det(A+I) = \pm 1$. It may be shown that there are only 2 such such matrices, up to conjugacy and inversion. Cf. [New: page 52]). Then A determines an orientation reversing homeomorphism of $S^1 \times S^1 \times S^1$. The fundamental group of the mapping torus of this homeomorphism is the HNN extension Z^3*_A, which is a PD_4-group of nonorientable type. It is easily seen to be a 3-knot group, but by Theorem 10 cannot be a 2-knot group. (Cappell and Shaneson used such matrices and mapping tori to construct PL 4-manifolds homotopy equivalent but not PL homeomorphic to RP^4 [CS 1976']).

If π is a PD_4-group and $\pi' = \pi''$, is M still aspherical? Is every (torsion free) 2-knot group π with $H^s(\pi;Z[\pi]) = 0$ for $s \leqslant 2$ a PD_4-group? Is every 3-knot group which is also a PD_4-group a 2-knot group?

Finally we shall show that any 2-knot whose group is a PD_4-group must be irreducible.

Theorem 11 *Let G be a PD_n-group over Q with $n > 2$. Then G is not a nontrivial free product with amalgamation over a cyclic subgroup.*

Proof If G is a nontrivial free product $G = A*_C B$ then A and B have infinite index in G. Therefore if G is a PD_n-group over Q the subgroups A and B have homological dimension over Q at most $n-1$, by [B: Proposition 9.22]. Moreover if C is cyclic then $h.d._Q C \leqslant 1$. A Mayer-Vietoris argument (as in [B: Theorem 2.10]) then shows that $h.d._Q G \leqslant \max\{n-1, 2\}$. Thus we must have $n = 2$. □

Corollary *If K is a 2-knot such that $M(K)$ is aspherical then K is not a nontrivial satellite knot. In particular, K is irreducible.*

Proof Let K_1 and K_2 be two 2-knots, and let γ be an element of πK_1. If γ has finite order let q be that order; otherwise let $q = 0$. Let w be a meridian in πK_2. Then by van Kampen's Theorem $\pi\Sigma(K_2;K_1,\gamma) = (\pi K_2/\langle\langle w^q\rangle\rangle)*_C\pi K_1$, where the amalgamation is over $C = Z/qZ$ and w is identified with γ in πK_1 [Ka 1983]. □

Chapter 4 THE RANK 1 CASE

It is a well known consequence of the asphericity of the complements of classical knots that classical knot groups are torsion free. This is also true of any 2-knot group which has a nontrivial torsion free abelian normal subgroup of rank at least 2, by Theorems 3 and 4 of Chapter 3. The first examples of higher dimensional knots whose groups have torsion were given by Mazur [Ma 1962] and Fox [Fo 1962]. Their examples have finite commutator subgroup, and hence in each of them some power of a meridian is a central element of infinite order. In this chapter we shall determine all the 2-knot groups with finite commutator subgroup, and we shall also consider the larger class of groups having abelian normal subgroups of rank 1. Most of the groups π with π' finite can be realized by twist spinning. This construction was introduced by Zeeman in order to study Mazur's example. Fox used his method of hyperplane cross sections, but his knots were later shown to be also twist spun knots [Ka 1983']. Fox also gave another striking example, with group Φ, which is certainly not even a fibred knot as Φ' is not finitely generated.

If π is a 2-knot group with an abelian normal subgroup A of rank 1 then either π' is finite ($e(\pi/A) = 0$) or π/zA is a PD_4^+-group over Q ($e(\pi/A = 1$) or π is an extension of Φ by a finite normal subgroup ($e(\pi/A) = 2$) or $e(\pi/A) = \infty$. After settling the case when π' is finite, we shall prove two general theorems. We first show that if A is not contained in π' and if moreover $e(\pi/A) = 1$ then π' is a PD_3^+-group. Next we show that if A is contained in π' and π' is finitely presentable then it is a PD_3^+-group with nontrivial centre. Finally we shall show that if $e(\pi/A) = 2$ then π must be Φ.

Cohomological periodicity

We saw in Chapter 2 that if the commutator subgroup of a 2-knot group π is finite, then all of its abelian subgroups are cyclic, and therefore π' has periodic cohomology [CE: page 262]. We shall establish this fact directly, in a stronger form, in our next theorem. (We shall use the full strength of the theorem in Chapter 7).

Theorem 1 *Let K be a 2-knot with group π such that π' is finite.*

Then $\widetilde{M}(K)$ *is homotopy equivalent to* S^3.

Proof Let C be an infinite cyclic central subgroup of π, and let M_C be the covering space of M with group C. Then M_C is a closed orientable 4-manifold with fundamental group Z, and $\chi(M_C) = [\pi:C]\chi(M) = 0$. The homology groups of $\widetilde{M}$ may be regarded as modules over the ring $\Gamma = Z[C] = Z[c,c^{-1}]$. By the Wang sequence for the projection of $\widetilde{M}$ onto M_C, multiplication by $c-1$ maps $H_2(\widetilde{M};Z)$ onto itself. But by equivariant Poincaré duality and the Universal Coefficient spectral sequence we have $H_2(\widetilde{M};Z) = \overline{Hom_\Gamma(H_2(\widetilde{M};Z),\Gamma)}$. Hence $H_2(\widetilde{M};Z) = 0$. Since $\pi_1(M_C)$ has two ends $H_3(\widetilde{M};Z) \cong Z$ and since $\widetilde{M}$ is an open 4-manifold $H_4(\widetilde{M};Z) = 0$. Therefore the map from S^3 to $\widetilde{M}$ representing a generator of $\pi_3(M)$ is a homotopy equivalence. □

Corollary *The commutator subgroup* π' *has cohomological period dividing* 4, *and the meridianal automorphism induces the identity on* $H_3(\pi';Z)$.

Proof The first assertion follows immediately from the Cartan-Leray spectral sequence for the projection p of $\widetilde{M} \sim S^3$ onto M' (cf. [CE: page 357]). By the Wang sequence for the projection of M' onto M we see that the meridianal automorphism induces the identity on $H_3(M';Z)$. As the spectral sequence also gives $H_3(\pi';Z) \cong Coker(H_3(p)) \cong Z/|\pi'|Z$ the second assertion is also immediate. □

We may use this corollary to give further examples of high dimensional knot groups which are not 2-knot groups. If p is a prime greater than 3 then $Z \times SL(2,p)$ is a 3-knot group with commutator subgroup the finite superperfect group $SL(2,p)$ (cf. the discussion of centres in Chapter 2). However as $SL(2,p)$ has cohomological period $p-1$ (if $p \equiv 1$ mod (4)) or $2(p-1)$ (if $p \equiv 3$ mod (4)) [LM 1978], it can only be the commutator subgroup of a 2-knot group if $p = 5$. The group $Z \times I^*$ is the group of the 5-twist spin of the trefoil knot; this was essentially the example found by Mazur [Ma 1962].

We shall follow the exposition of Plotnick and Suciu [PS 1987] (rather than the original one of [Hi 1977]) in determining which finite groups with cohomological period 4 have meridianal automorphisms. Let $Q(1)$ be the quaternion group, which has a presentation $\langle x,y \mid x^2 = (xy)^2 = y^2\rangle$, and let σ be the automorphism which sends x and y to y and xy respectively. For each $k \geqslant 1$ let $T(k)$ be the group with presentation

$$\langle x,y,z \mid x^2 = (xy)^2 = y^2,\ zxz^{-1} = y,\ zyz^{-1} = xy,\ z^{3^k} = 1\rangle.$$

(Thus $T(k)' \cong Q(1)$, and $T(k)$ is a semidirect product $Q(1)\times_\sigma Z/3^k Z$). The binary icosahedral group I^* has a presentation $\langle x,y \mid x^2 = (xy)^3 = y^5\rangle$.

Theorem 2 *Let π be a 2-knot group with π' finite. Then $\pi' \cong P\times Z/nZ$ where $P \cong 1$, $Q(1)$, $T(k)$ or I^* and $(n,2|P|) = 1$.*

Proof All the finite groups with the property that every abelian subgroup is cyclic are listed, in 6 classes, on pages 179 and 195 of [Wo]. As the meridianal automorphism of the commutator subgroup of a knot group induces a meridianal automorphism on the quotient by any characteristic subgroup, we may dismiss from consideration those which have abelianization cyclic of even order. There remain types I,II,III and V.

Each group of type I is metacyclic, with presentation $\langle a,b \mid a^m = b^n = 1,\ bab^{-1} = a^r\rangle$, where $(m,n(r-1)) = 1$ and $r^n \equiv 1 \bmod (m)$. A meridional automorphism σ must induce a meridianal automorphism on the abelianization. Therefore $\sigma(a) = a^i$ where $(i,m) = 1$ and $\sigma(b) = a^j b^k$ where $(k,n) = (k-1,n) = 1$. The condition that $\sigma(bab^{-1}) = \sigma(a^r)$ then implies that $ir^k \equiv ir \bmod (m)$, and so $r = 1$. Therefore the group is cyclic of odd order.

Each group of type II has a subgroup of type I of index 2, and has a presentation

$$\langle a,b,c \mid \text{as in } I;\ \text{also } c^2 = b^{n/2},\ cac^{-1} = a^p,\ cbc^{-1} = b^q\rangle$$

where $p^2 \equiv r^{q-1} \equiv 1 \bmod (m)$, $n = 2^u \nu$ for some $u \geqslant 2$ and odd ν, $q+1 \equiv 0 \bmod 2^u$ and $q^2 \equiv 1 \bmod (n)$.

The conditions $(m,n(r-1)) = 1$ and $r^n \equiv 1 \bmod (m)$ imply that m is odd and $(m,n) = 1$. The subgroup generated by $\{a,b^{2^u}\}$ is the unique maximal subgroup of odd order, and so is characteristic. Therefore a meridianal automorphism on such a group induces a meridianal automorphism on the quotient, which has a presentation

$$<b,c \mid b^{2^u} = 1,\ c^2 = b^{2^{u-1}},\ cbc^{-1} = b^q>.$$

If $u > 2$ the only elements of order 2^u in this quotient are the odd powers of b, and so no automorphism can be meridianal. Therefore we may assume that $u = 2$.

The subgroup generated by a is also characteristic. Any meridianal automorphism of the quotient by this subgroup must map b to $b^e c$ for some e and so must map b^4 to $b^{2e(q+1)}$. Therefore $(q+1,\nu) = 1$ and the congruences above then imply that $q \equiv 1 \bmod (\nu)$. Hence the quotient is isomorphic to $Q(1) \times Z/\nu Z$.

Thus a meridianal automorphism of a group of type II with such a presentation must map a, b^4, b^ν and c to a^f, $a^g b^{4e}$, $a^h b^\nu c$ and $a^i b^\nu$ respectively. Since it must preserve the relations $b^\nu a b^{-\nu} = a^r$, $b^4 a b^{-4} = a^r$ and $cac^{-1} = a^p$, we obtain the further congruences $fpr^\nu \equiv fr^\nu$, $fr^{4e} \equiv fr^4$ and $fr^\nu \equiv fp \bmod (m)$. As moreover (f,m) must be 1 and $p^2 \equiv 1 \bmod (m)$ and $(4(e-1),\nu) = 1$, we find that $r \equiv p \equiv 1 \bmod (m)$. Therefore the group is isomorphic to $Q(1) \times Z/m\nu Z$, and $m\nu$ is odd.

Each group of type III is an extension of a group of type I by $Q(1)$, and has a presentation

$$<a,b,x,y \mid \text{as in } I;\ \text{also } x^2 = (xy)^2 = y^2,\ ax = xa,\ ay = ya,$$
$$bxb^{-1} = y,\ byb^{-1} = xy>$$

where the order n of b is an odd multiple of 3. Since the Sylow 2-subgroup $Q(1)$ is characteristic, the quotient admits a meridianal automorphism; since it is of type I it must be cyclic, and so $r = 1$. Let $n = 3^k s$ where s is not divisible by 3. Then the subgroup generated by ab^3 is central and of order ms. The subgroup generated by $\{x,y,b^s\}$ is isomorphic to

$T(k)$, and so the whole group is isomorphic to $T(k)\times Z/msZ$, where $(ms,6) = 1$.

The only groups of type V that we need consider are direct products $J\times SL(2,p)$ where J is of type I, p is a prime greater than 3 and $(|J|,|SL(2,p)|) = 1$. If such a direct product (with factors of coprime order) admits a meridianal automorphism, then so do its factors. Therefore J is cyclic of odd order. As remarked above, the cohomological dimension of $SL(2,p)$ is greater than 4 if $p > 5$. This completes the theorem. □

It is well known that each such group is the fundamental group of a 3-dimensional spherical space form [Wo]. In particular it has trivial second homology. Therefore each meridianal automorphism of such a group can be realized by some 3-knot group. Plotnick and Suciu study n-knots which are fibred with fibre a punctured spherical space form, and show that if $n > 2$ then π' must be cyclic [PS 1987].

Meridianal automorphisms

Having found the groups with cohomological period 4 which admit *some* meridianal automorphism, we must next determine all such automorphisms (up to inversion and conjugacy in the outer automorphism group). We shall show that in fact there is in each case just one 2-knot group with a given finite commutator subgroup. Throughout this section π shall be a 2-knot group with finite commutator subgroup.

Our results shall be developed in a number of lemmas and then summarized in a theorem. The first lemma is self-evident.

Lemma 1 *If a group $G = H\times J$ with $(|H|,|J|) = 1$ then an automorphism ϕ of G corresponds to a pair of automorphisms ϕ_H and ϕ_J of H and J respectively, and ϕ is meridianal if and only if ϕ_H and ϕ_J are.*
□

Lemma 2 *The meridianal automorphism of a cyclic direct factor of π' is the involution.*

Proof The endomorphism $[s]:x\rightarrow x^s$ of the cyclic group of order m is a meridianal automorphism if and only if $(s-1,m) = (s,m) = 1$. If the group is a direct factor of π' then it is a direct summand of $\pi'/\pi'' = H_1(M(K);\Lambda)$ and so Theorem 3 of Chapter 2 implies that $s^2 \equiv 1 \bmod (m)$. Hence we must have $s \equiv -1 \bmod (m)$. □

Lemma 3 *An automorphism of $Q(1)$ is meridianal if and only if its image in $Out(Q(1))$ equals that of σ or σ^{-1}.*

Proof It is easy to see that an automorphism of $Q(1)$ induces the identity on $Q(1)/Q(1)'$ if and only if it is inner, and that every automorphism of $Q(1)/Q(1)'$ lifts to one of $Q(1)$. Therefore $Out(Q(1)) = Aut(Q(1)/Q(1)')$. Moreover as $Q(1)$ is solvable an automorphism is meridianal if and only if the induced automorphism of $Q(1)/Q(1)'$ is meridianal. The latter are represented by the images of σ and σ^{-1} in $Out(Q(1))$. □

A more detailed calculation shows that every meridianal automorphism of $Q(1)$ is conjugate to σ or σ^{-1} by an inner automorphism.

Lemma 4 *All nontrivial automorphisms of I^* are meridianal. Moreover each automorphism is conjugate to its inverse, and $Out(I^*) = Z/2Z$.*

Proof An elementary calculation shows that if an automorphism of a group G induces the identity on $G/\zeta G$, then it is the identity on all commutators. Therefore if G is a perfect group the natural map from $Aut(G)$ to $Aut(G/\zeta G)$ is injective. Since the only nontrivial proper normal subgroup of I^* is its centre ($\zeta I^* = Z/2Z$) the first assertion is immediate. The general linear group $GL(2,5)$ acts on $I^* = SL(2,5)$ by conjugation, and the kernel of this action is the subgroup of scalar matrices. Therefore there is a monomorphism from $PGL(2,5)$ to $Aut(I^*)$. Now $I^*/\zeta I^* = PSL(2,5)$ is simple and has order 60, and so is isomorphic to A_5. Since $Aut(A_5) = S_5$ and $|PGL(2,5)| = |S_5| = 120$, it follows that $Aut(I^*) \cong S_5$, and $Out(I^*)$ has order 2. Since the conjugacy class of a permutation is determined by its cycle structure, each automorphism is conjugate to its inverse. □

The argument identifying $Aut(I^*)$ with S_5 is taken from [Pl 1983], as is the next lemma.

Lemma 5 [Pl 1983] *The nontrivial outer automorphism class of* I^* *cannot be realized by a 2-knot group.*

Proof Let ω be the automorphism of I^* represented by the matrix $\begin{pmatrix} 2 & 0 \\ 0 & 1 \end{pmatrix}$ in $GL(2,5)$. Then ω represents the nontrival outer automorphism class. We shall show that the induced automorphism ω_3 of $H_3(I^*;Z)$ is not the identity. (Plotnick shows that it is in fact multiplication by 49). It shall suffice for us to consider the effect on the 5-torsion. As the inclusion of the Sylow subgroups induces iso norphisms of the corresponding torsion in the homology, the effect of α on the 5-torsion in the homology is detected by its action on a Sylow 5-subgroup.

The matrix $\gamma = \begin{pmatrix} 1 & 1 \\ 0 & 1 \end{pmatrix}$ generates a Sylow 5-subgroup of I^*, and $\omega\gamma\omega^{-1} = \gamma^2$. Therefore the induced map on $H^2(Z/5Z;Z) = H_1(Z/5Z;Z) = Z/5Z$ is multiplication by 2. Since the square of a generator for $H^2(Z/5Z;Z)$ generates $H^4(Z/5Z;Z) = H_3(Z/5Z;Z)$, we see that ω_3 is multiplication by 4 on 5-torsion. Since inner automorphisms induce the identity on homology, it now follows from the corollary to Theorem 1 that the meridional automorphism of such a 2-knot group must be inner. □

The elements of order 2 in $A_5 \cong Inn(I^*)$ are all conjugate, as are the elements of order 3. However there are two conjugacy classes of 5-cycles in A_5.

Lemma 6 *The meridianal automorphisms of* $T(k)$ *which are realizable by a 2-knot form two conjugacy classes in* $Aut(T(k))$, *and have the same image in* $Out(T(k))$.

Proof Let ρ be the automorphism of $T(k)$ which sends x, y and z to y^{-1}, x^{-1} and z^{-1} respectively. Then the image of ρ under the natural map α from $Aut(T(k))$ to $Aut(T(k)/T(k)') \cong (Z/3^k Z)^*$ is [2], which generates $(Z/3^k Z)^*$. The kernel of α clearly contains the inner automorphisms, and it is not hard to check that conversely any automorphism which induces the identity on the abelianization is inner. Thus $Out(T(k)) = (Z/3^k Z)^*$.

Let ξ, η and ζ be the inner automorphisms determined by conjugation by x, y and z respectively (i.e. $\xi(g) = xgx^{-1}$ and so on). Then $Aut(T(k))$ is generated by ρ, ξ, η and ζ. Moreover $\zeta^3 = 1$ and $\xi = \zeta^{-1}\eta\zeta$, and it follows that $Aut(T(k))$ has a presentation

$$<\rho,\eta,\zeta \mid \rho^{2.3^{k-1}} = \eta^2 = \zeta^3 = [\eta,[\eta,\zeta]] = 1,\ \rho\zeta\rho^{-1} = \zeta^2,$$
$$\rho\eta\rho^{-1} = [\eta,\zeta] = \zeta^{-1}\eta\zeta>$$

Since $T(k)$ is solvable, an automorphism is meridianal if and only if the induced automorphism of $T(k)/T(k)'$ is meridianal. Any such automorphism is conjugate to either ρ^{2j+1} or to $\rho^{2j+1}\eta$ for some $0 \leqslant j < 3^{k-1}$. (Note that 3 divides $2^{2j}-1$ but does not divide $2^{2j+1}-1$). However among them only those with $2j+1 = 3^{k-1}$ satisfy the isometry condition of Theorem 3 of Chapter 2. □

Note that $\rho^{3^{k-1}}$ is the involution which sends x, y and z to y^{-1}, x^{-1} and z^{-1} respectively, while the automorphisms in the other meridianal conjugacy class have order 4. Since $Out(T(k))$ is cyclic, conjugate meridianal automorphisms are in fact conjugate by inner automorphisms of $T(k)$.

If we take these lemmas and Theorem 2 together we obtain the following result.

Theorem 3 *Let π be a 2-knot group with π' finite. Then $\pi' \cong P \times Z/nZ$ where $P \cong 1$, $Q(1)$, $T(k)$ or I^*, and the meridianal automorphism is -1 on the cyclic factor, is conjugation by a noncentral element on I^*, and sends x, y in $Q(1)$ to y, xy and x, y, z in $T(k)$ y^{-1}, x^{-1} and z^{-1} respectively.* □

Note that when $P = I^*$ there is an isomorphism $\pi \cong I^* \times (\pi/I^*)$. Yoshikawa has shown that all the groups allowed by this theorem can be realized by fibred 2-knots [Yo 1980]. The commutator subgroups of the 2-, 3-, 4- and 5- twist spins of the trefoil knot are $Z/3Z$, $Q(1)$, $T(1)$ and I^* repectively. The other cyclic groups are realized by the 2-twist spins of the other 2-bridge knots. The direct products of $T(k)$ and I^* with cyclic groups are realized by the 2-twist spins of certain pretzel knots. The remaining groups $Q(1)\times Z/nZ$ cannot be realized by twist spins (cf. Chapter 5); are they realizable by (fibred) 2-knots which are smooth in the standard smooth structure on S^4? (This is so when $n = 5$ or 11 [Ka 1988]).

Plotnick and Suciu show that if $P = 1$, $Q(1)$, $T(k)$ or I^* then π has 1, 1, 2 or 4 weight orbits respectively. (This can be proven using the criterion of Theorem 8 of Chapter 2). They also show that if $\pi' = Q(1)$, $T(1)$ or I^* then each weight orbit is realized by a branched twist spin of a torus knot [PS 1987].

Kanenobu has shown that for every n there is a 2-knot group which has an element of order exactly n [Ka 1980].

Infinite commutator subgroup

We may henceforth assume that if π is a 2-knot group then π' is infinite. We shall however continue to assume that π has a torsion free abelian normal subgroup of rank 1. In some circumstances we do not need to assume *a priori* that the subgroup be torsion free.

Lemma 7 *Let J be a finitely generated group with $J/J' \cong Z$ and A an abelian normal subgroup of rank 1 which is not contained in J'. Then J has a central element of infinite order which is not in J'.*

Proof Let $x_1, \cdots, x_n$ be a set of generators for J and let j be an element of A which is not in J'. Since A is normal, each commutator $[j,x_i]$ is in $A \cap J'$ and therefore has finite order, e_i say. Let $e = \Pi e_i$. Then j^e commutes with all the generators and so is central, and has infinite order modulo J'. □

Theorem 4 *Let π be a 2-knot group with an abelian normal subgroup* *of rank* 1 *which is not contained in* π'. *Then* π' *is finitely presentable, and* $\zeta\pi$ *is not contained in* π'. *Moreover* π *is a* PD_4^+*-group if and only if* π' *is a* PD_3^+*-group if and only if* π' *has* 1 *end.*

Proof By Lemma 7 there is an infinite cyclic central subgroup A which is not contained in π'. The subgroup generated by $A\cup\pi'$ has finite index in π and is isomorphic to $A\times\pi'$, so π' is finitely presentable. If π is a PD_4^+-group then so is $A\times\pi'$. Hence π' is a PD_3^+-group and so $e(\pi') = 1$. Since $A\cap\pi' = 1$, $(\pi/A)' = \pi'$ and has finite index in π/A, so $e(\pi') = 1$ implies that $e(\pi/A) = 1$, and then $H^s(\pi;Z[\pi]) = 0$ for $s \leqslant 2$ by Theorem 5 of Chapter 3. Corollary 1 of Theorem 3 of Chapter 3 then implies that π is a PD_4^+-group. □

The group of a torus knot has centre Z with $\zeta\pi\cap\pi' = 1$ and π' is free of even rank (so $e(\pi') = \infty$). The group of a twist spun prime knot (other than a torus knot or certain rational knots) is a PD_4^+-group with centre Z not contained in its commutator subgroup. (The group of the r-twist spin of the (p,q)-torus knot has centre of rank 2, except when $p^{-1}+q^{-1}+r^{-1} > 1$. See Chapter 5). In Chapter 7 we shall show that under the assumptions of Theorem 4 the infinite cyclic cover M' is an orientable PD_3-complex. In particular, if π is torsion free then π' is a free product of PD_3^+-groups and free groups.

Lemma 8 *Let* J *be a group with* $J/J' \cong Z$ *and suppose that* j *is an element of infinite order in* $\zeta J'$. *Then* J *has a nontrivial torsion free abelian normal subgroup.*

Proof We note first that j commutes with its conjugates. Therefore $C = \langle\langle j\rangle\rangle_J$ is an abelian group. Moreover it is a module over $Z[J/J'] \cong \Lambda$, and is cyclic as a module. Therefore the Z-torsion submodule zC is finitely generated and so has finite exponent, e say. The subgroup $A = \langle\langle j^e\rangle\rangle_J$ is then a nontrivial torsion free abelian normal subgroup of J. □

Lemma 9 *Let J be a group with J' finitely generated and $J/J' \cong Z$ and suppose that J has an abelian normal subgroup of rank 1 which is contained in J'. Then J has a torsion free abelian normal subgroup of rank 1.*

Proof We may suppose that J is generated by $t, x_1, \cdots, x_n$ where the image of t generates the abelianization, and where the other generators generate J'. Let j be an element of infinite order which is contained in an abelian normal subgroup A of rank 1. Since A is normal, each commutator $[j,x_i]$ is in A; since A/zA is torsion free and of rank 1, $Aut(A/zA)$ is abelian and so each $[j,x_i]$ is in zA. Let e_i be the order of $[j,x_i]$ and let $e = \Pi x_i$. Then j^e commutes with all the generators and so we may apply Lemma 8 to conclude that A contains a nontrivial torsion free abelian subgroup which is normal in J and necessarily of rank 1. □

Theorem 5 *Let π be a 2-knot group with π' finitely generated and with an abelian normal subgroup of rank 1, which is contained in π'. Then π is a PD_4^+-group and π' has nontrivial centre. If moreover π' is almost finitely presentable (FP_2) then it is a PD_3^+-group.*

Proof By Lemma 9 we may assume that there is a torsion free abelian normal subgroup A of rank 1 in π'. Since A is torsion free and of rank 1, $Aut(A)$ is abelian and so A is central in π'. If π'/A is finite then π is a finite extension of Φ, by Theorem 5 of Chapter 3. Therefore if π' is finitely generated then $(\pi/A)'$ is finitely generated and infinite, so $e(\pi/A) = 1$, and π is a PD_4^+-group as in the previous theorem.

Since $c.d.\pi' = 3$ the augmentation $Z[\pi']$-module has a projective resolution

$$0 \to P_3 \to P_2 \to P_1 \to P_0 \to Z \to 0.$$

If π' is FP_2 then we may assume that the modules P_i are finitely generated for $i \leqslant 2$, and the natural map from $H^s(\pi';Z[\pi'])\otimes\Lambda$ to $H^s(\pi';Z[\pi])$ is an isomorphism for $s \leqslant 2$ (cf. [B: Theorem 5.3]). This is also true for $s = 3$. Let C be an infinite cyclic subgroup of $\zeta\pi'$ and let $D = \pi'/C$.

Then the LHS spectral sequence for π' as an extension of D by C gives an isomorphism $H^3(\pi';Z[\pi']) = H^2(D;H^1(C;Z[\pi'])) = H^2(D;Z[D])$, since C is finitely generated. Likewise, writing $Z[\pi] = Z[\pi']^{(Z)}$, we have $H^3(\pi';Z[\pi]) = H^2(D;Z[D]^{(Z)}) = H^2(D;Z[D])^{(Z)}$ since D is FP_2. Therefore $H^3(\pi';Z[\pi]) = H^3(\pi';Z[\pi'])^{(Z)}$. On keeping track of the direct sum decompositions, we see that in fact $H^3(\pi';Z[\pi]) = H^3(\pi';Z[\pi'])\otimes\Lambda$ as a Λ-module.

The LHS spectral sequence for π as an extension of Z by π', with coefficients $Z[\pi]$, reduces to a Wang sequence

$$\cdots \to H^q(\pi';Z[\pi]) \to H^q(\pi';Z[\pi]) \to H^{q-1}(\pi;Z[\pi]) \to \cdots$$

Using the above information on $H^q(\pi';Z[\pi])$ we find that $H^q(\pi';Z[\pi']) = H^{q+1}(\pi;Z[\pi]) = 0$ for $q \neq 3$ and $H^3(\pi';Z[\pi']) = H^4(\pi;Z[\pi]) = Z$. Thus if we dualize the above $Z[\pi']$-resolution of Z by means of $P^* = Hom_{Z[\pi']}(P,Z[\pi'])$ we get an exact sequence

$$0 \to P_0^* \to P_1^* \to P_2^* \to P_3^* \to H^3(\pi';Z[\pi']) = Z \to 0.$$

The dual of a projective module P is finitely generated if and only if P is. Therefore P_3^* and hence P_3 are finitely generated. Thus π' is FP_3. As $H^q(\pi';Z[\pi']) = Z$ if $q = 3$ and is 0 otherwise, π' is a PD_3^+-group. □

In Chapter 5 we shall show that if a PD_3^+-group has a subgroup of finite index with nontrivial centre which maps onto Z then it is the fundamental group of a Seifert fibred 3-manifold. In this case the centre must be finitely generated. There are fibred knots with such groups, even with π a poly-Z-group, as we shall see in Chapter 6. Yoshikawa has constructed a 2-knot whose group has centre Z contained in π' and for which $\pi/\zeta\pi$ has infinitely many ends [Yo 1982]. The only known 2-knot group with an abelian normal subgroup which is not finitely generated is the group Φ.

The group Φ

If π is a 2-knot group with an abelian normal subgroup A of rank 1 and if $e(\pi/A) = 2$ then by Theorem 5 of Chapter 3 π has a finite normal subgroup, N say, with $\pi/N \cong \Phi$. In this section we shall examine this case more closely, and we shall see that N must be trivial.

For each $m \geqslant 1$ let $\Phi(m)$ be the group with presentation $\langle a,t \mid tat^{-1} = a^{2m} \rangle$. Then every subgroup of finite index in $\Phi(1) = \Phi$ is isomorphic to $\Phi(m)$ for some m. (For if Θ is such a subgroup then $\Theta \cap \Phi' \cong \lambda Z[\frac{1}{2}]$ for some odd $\lambda \geqslant 1$, so Θ is generated by a^{λ} and $t^m a^{\mu}$, for some $m \geqslant 1$ and μ in $Z[\frac{1}{2}]$, with a single relation $(t^m a^{\mu})a^{\lambda}(t^m a^{\mu})^{-1} = a^{\lambda 2m}$). Let $\hat{\Phi}(m)$ be the kernel of the homomorphism from $\Phi(m)$ onto Z sending a to 0 and t to 1. (As an abelian group $\hat{\Phi}(m) \cong \Phi' \cong Z[\frac{1}{2}]$).

Lemma 10 *Let Y be a closed orientable 4-manifold with $\chi(Y) = 0$ and such that there is an epimorphism f from $G = \pi_1(Y)$ to $\Phi(m)$ for some m, with finite kernel. Then the integral homology groups of the infinite cyclic covering space $\hat{Y}$ determined by $\hat{G} = f^{-1}(\hat{\Phi}(m))$ are finitely generated Λ-torsion modules, and $H_2(\hat{G};Z)$ is finite cyclic of odd order.*

Proof Since Y is compact and Λ is noetherian, the groups $H_i(\hat{Y};Z) = H_i(Y;\Lambda)$ are finitely generated as Λ-modules. Since Y is orientable, $\chi(Y) = 0$ and $H_1(Y;Z)$ has rank 1, $H_2(Y;Z)$ is finite and the rest of the first assertion follows from the Wang sequence for the projection of $\hat{Y}$ onto Y. By Poincaré duality and Hopf's theorem $H_2(\hat{G};Z)$ is a quotient of $Ext^1_{\Lambda}(\hat{G}/\hat{G}',\Lambda)$. (Cf. Theorem 3 of Chapter 2). By assumption there is an exact sequence $0 \to T \to \hat{G}/\hat{G}' \to \hat{\Phi}(m) \cong \Lambda/(t-2^m) \to 0$ where T is a finite Λ-module. Therefore $Ext^1_{\Lambda}(\hat{G}/\hat{G}',\Lambda) \cong \Lambda/(t-2^m)$ and so $H_2(\hat{G};Z)$ is a quotient of $\Lambda/(2^m t-1)$, which is isomorphic to $Z[\frac{1}{2}]$ as an abelian group. Now $\hat{G}/ker(f) \cong Z[\frac{1}{2}]$ also, and $H_2(Z[\frac{1}{2}];Z) = Z[\frac{1}{2}] \wedge Z[\frac{1}{2}]$ (by [R: page 334]) $= 0$, and so by the LHS spectral sequence for $\hat{G}$ as an extension of $Z[\frac{1}{2}]$ by $ker(f)$ we see that $H_2(\hat{G};Z)$ is finite. But a finite quotient of $Z[\frac{1}{2}]$ is cyclic of odd order. □

Lemma 11 *Let A be an abelian group and F a field of characteristic $\neq 2$, considered as a trivial A-module. Then the map induced by cup product from $H^1(A;F) \wedge H^1(A;F)$ to $H^2(A;F)$ is injective.*

Proof We may identify $H^1(A;F)$ with $Hom(A,F)$ and $H^2(A;F)$ with $\{w:A^2 \to F \mid \partial^2 w = 0\}/\{\partial^1 f \mid f:A \to F\}$ where $\partial^1 f(a,b) = f(a)+f(b)-f(ab)$ and $\partial^2 w(a,b,c) = w(b,c)-w(ab,c)+w(a,bc)-w(a,b)$ for all a, b, c in A and set maps f and w. The cup product of two elements f and g in $H^1(A;F)$ is represented by the function sending (a,b) in A^2 to $f(a)g(b)$ in F.

There is a natural monomorphism μ from $H^1(A;F) \wedge H^1(A;F)$ to $\wedge^2(A,F)$ (the space of skew symmetric bilinear maps from A^2 to F) given by $\mu(\Sigma f_i \wedge g_i)(a,b) = \Sigma(f_i(a)g_i(b)-f_i(b)g_i(a))$. Suppose that f_i, g_i in $H^1(A;F)$ are such that $\Sigma f_i \cup g_i = 0$ in $H^2(A;F)$. Then there is a function h from A to F such that $\Sigma f_i(a)g_i(b) = h(a)+h(b)-h(ab)$ for all a, b in A. Then $\mu(\Sigma f_i \wedge g_i)(a,b) = h(ba)-h(ab) = 0$ for all a, b in A , since A is abelian. Since μ is a monomorphism, $\Sigma f_i \wedge g_i$ must be 0. □

For more general groups G and for coefficients including fields of characteristic 2 or the ring Z the kernel of cup product may be related to the subquotients of descending central series for G [Hi 1987].

Lemma 12 *Let π be a 2-knot group which is an extension of Φ by a finite normal subgroup N. Then $N \cong Q(1)$ or is trivial.*

Proof Let K be a 2-knot with such a group π, and fix a meridian t in π. Let A be a maximal abelian subgroup of N and let $G_1 = C_\pi(A)$. Then since N is finite and normal in π the index $m = [\pi:G_1]$ is finite. In particular t^m is in G_1 and G_1 maps onto Z, with kernel $\hat{G}_1$ say. Let G be the subgroup of π generated by $\zeta\hat{G}_1$ and t^m. Then G has finite index in π and so there is an epimorphism f from G onto $\Phi(m)$, with kernel A. Moreover $\hat{G} = f^{-1}(\hat{\Phi}(m))$ is abelian, and is an extension of $\hat{\Phi}(m) \cong Z[½]$ by the finite abelian group A, and so is isomorphic to $A \oplus Z[½]$, by [R: page 106]. Now $H_2(\hat{G};Z)$ is cyclic of odd order, by Lemma 10. On the other hand, by [R: page 334] it is isomorphic to $A \wedge A \oplus (A \otimes Z[½])$. Therefore A is cyclic.

Now let M_G and $\hat{M}_G$ be the covering spaces of $M(K)$ determined by the subgroups G and $\hat{G}$ of π, and let $F = F_p$, where p is an odd prime. Then the classifying map from $\hat{M}_G$ to $K(\hat{G},1)$ is 2-connected, so $H^1(\hat{G};F) \cong H^1(\hat{M}_G;F)$ and $H^2(\hat{G};F)$ maps injectively to $H^2(\hat{M}_G;F)$. If p divides the order of A then $H^1(\hat{G};F) \cong Hom(A,F) \oplus Hom(Z[½],F) \cong F^2$. The image of cup product in $H^2(\hat{M}_G;F)$ is then 1-dimensional (since it is nontrivial, by Lemma 11), and so is nontrivially paired with some element of $H^1(\hat{G};F)$, by Milnor duality for $\hat{M}_G$ as an infinite cyclic cover of M_G. But there can be no nontrivial alternating trilinear form on a 2-dimensional vector space. Therefore A must be a 2-group.

Since every abelian subgroup of N is a cyclic 2-group N must be itself a cyclic 2-group, or a generalized quaternion group $Q(n)$, with presentation $<x,y \mid x^2 = y^{2^n}, x^4 = 1>$ for some n [R: 5.3.6]. Suppose that $N \neq Q(1)$. Then N has no automorphism of odd order and so, returning to the knot group π, we must have $\pi' \cong N \times \Phi'$. Therefore N must admit a meridional automorphism. But this is impossible if N is cyclic of even order, or if $N \cong Q(n)$ for some $n > 1$. Thus $N \cong Q(1)$ or is trivial. □

Theorem 6 *Let π be a 2-knot group which has an abelian normal subgroup A of rank 1 such that π/A has 2 ends. Then $\pi \cong \Phi$.*

Proof We shall assume that π is an extension of Φ by $Q(1)$ and use equivariant Poincaré duality with coefficients $\Xi = F_2[\Phi]$ to deduce a contradiction. We shall first describe some of the properties of this noncommutative ring. Since Φ is a torsion free 1-relator group it has cohomological dimension 2 and so the ring Ξ has global dimension 2. A presentation for the augmentation module F_2 may be obtained by means of the free differential calculus; there is an exact sequence of left Ξ-modules

$$0 \to \Xi \xrightarrow{\partial_2} \Xi^2 \xrightarrow{\partial_1} \Xi \xrightarrow{\varepsilon} F_2 \to 0$$

where $\varepsilon(g) = 1$ for all g in Φ, $\partial_1(\theta,\phi) = \theta(a-1)+\phi(t-1)$ and $\partial_2(\xi) = (\xi(t+a+1), \xi(a^2+1))$. As a group ring Ξ has a natural involution, defined by $\bar{g} = g^{-1}$ for all g in Φ. Moreover Ξ is a twisted polynomial ring. Let $D = F_2[a^{2^{-\infty}}] = F_2[a_n \mid n \text{ in } Z]/(a_{n+1}-a_n^2 \mid n \text{ in } Z)$. This is a commutative Bezout domain (i.e. finitely generated ideals are principal) and its field of fractions $E = F_2(a^{2^{-\infty}})$ is perfect, i.e. the squaring map σ is an automorphism. Then $\Xi = D[a^{-1}][t,t^{-1};\sigma]$: each element may be expressed uniquely as a sum $\Sigma t^m p_m(a)$ over m in a finite subset of Z, where each $p_m(a)$ is in $D[a^{-1}]$, while the multiplication is determined by $a_{n+1}t = ta_n$ for all n , i.e. $tp(a) = \sigma(p(a))t$ for $p(a)$ in $D[a^{-1}]$. The ring Ξ has a skew field of fractions L, which as a right Ξ-module is the direct limit of the system $\{\Xi_\theta \mid 0 \neq \theta \ in\ \Xi\}$ where each $\Xi_\theta = \Xi$, the index set is ordered by right divisibility $(\theta \leqslant \phi\theta)$ and the map from Ξ_θ to $\Xi_{\phi\theta}$ sends ξ to $\phi\xi$, and so L is flat. (Note also that L contains the ring $E[t,t^{-1};\sigma]$ which is a skew Euclidean domain. See [C] for a full treatment of fields of fractions of twisted polynomial rings).

Let $M_{Q(1)}$ be the covering space of $M(K)$ with group $Q(1)$ and let C_* be the equivariant cellular chain complex of $M_{Q(1)}$ with coefficients F_2. For brevity, let H_p denote $H_p(M;\Xi) = H_p(C_*) = H_p(M_{Q(1)};F_2)$. Since $M_{Q(1)}$ is a connected open 4-manifold $H_0 = F_2$ (the augmentation Ξ-module) and $H_4 = 0$. The Ξ-module structure on $H_1 = Q(1)/Q(1)' = F_2^2$ is determined by a homomorphism from Φ to $Aut(Q(1)/Q(1)') = S_3$. Since the abelianization of π is infinite cyclic, the image of Φ cannot be trivial or of order 2, and there are essentially just two possibilities.

In the first case the image of Φ is cyclic of order 3, so $a-1$ and t^3-1 act as 0, but $t \neq 1$. It follows that $t-1$ is an automorphism, and that H_1 with this module structure, which we shall denote H_c, is simple, and the annihilator of a generator is the left ideal generated by $a-1$ and t^2+t+1. There is an exact sequence

$$0 \to \Xi \xrightarrow{-\partial_2^c} \Xi^2 \xrightarrow{-\partial_1^c} \Xi \to H_c \to 0$$

where $\partial_1^c(\theta,\phi) = \theta(a-1)+\phi(t^2+t+1)$ and $\partial_2^c(\xi) = (\xi(t^2+t(a-1)+(a-1)^3),\xi(a^4-1))$. Exactness at three of the modules, and $\partial_1^c\partial_2^c = 0$ are easily verified. We may show that $ker\,\partial_1^c = im\,\partial_2^c$ as follows. Suppose that $\partial_1^c(\theta,\phi) = 0$. Then $\theta(a-1) = \phi(t^2+t+1)$. We may write ϕ as $\phi = (\Sigma t^m p_m(a))(a-1)^{4d}$ for some $d \geqslant 0$ in $Z[½]$ and where $p_m(1) \neq 0$ for some m. Then $\phi(t^2+t+1) =$

$(\Sigma t^m P_m(a))(t^2+t(a-1)^d+(a-1)^{3d})(a-1)^d$. Since Ξ is a domain we may cancel factors from an equation. Thus if $d < 1$ we have $(\Sigma t^m p_m(a))(t^2+t(a-1)^d+(a-1)^{3d}) = \theta(a-1)^{1-d}$, so on substituting $a = 1$ we get $\Sigma t^m p_m(1) = 0$ in $F_2[t,t^{-1}]$. This contradicts the assumption on the polynomials $p_m(a)$. Thus $d \geqslant 1$, so we may write $\phi = \eta(a-1)^4$, and it follows that $\theta = \eta(t^2+t(a-1)+(a-1)^3)$.

Otherwise Φ maps onto $Aut(Q(1)/Q(1)')$, so t maps to an automorphism of order 2 and a maps to an automorphism of order 3, and the module (now denoted H_f) is again simple. There is an exact sequence

$$0 \to \Xi \xrightarrow{-\partial_2^f} \Xi^2 \xrightarrow{-\partial_1^f} \Xi \to H_f \to 0$$

where $\partial_1^f(\theta,\phi) = \theta(t+a)+\phi(a^2+a+1)$ and $\partial_2^f(\xi) = (\xi(a^4+a^2+1),\xi(t+a^3+a^2+a))$. (Exactness of the sequence is proved in a similar fashion).

For any left Ξ-module H, let $e^qH = Ext^q_{\Xi}(H,\Xi)$ and let H be the right Ξ-module with the conjugate action. From the three resolutions given above we may compute that $e^1F_2 = e^1H_c = e^1H_f = 0$. (In fact $e^0F_2 = e^1F_2 = 0$, which is equivalent to the fact that the group Φ has one end). Then Poincaré duality and the Universal Coefficient spectral sequence give $H_3 = 0$ in either case, and an exact sequence

$$\cdots \to e^0H_2 \to e^2H_1 \to H_1 \to e^1H_2 \to 0 .$$

Now since the skew field of fractions L is flat as a right module, $H_p(L\otimes_{\Xi}C_*) = L\otimes_{\Xi}H_p$, and so is nonzero only if $p = 2$. But since M has Euler characteristic 0, which is also the Euler characteristic of $L\otimes_{\Xi}C_*$ and therefore of $L\otimes_{\Xi}H_*$, we may conclude that $L\otimes_{\Xi}H_2 = 0$ also. Therefore $e^0H_2 = 0$ (since $e^0H_2 = Hom(H_2,\Xi)$ is contained in $Hom(L\otimes_{\Xi}H_2,L)$) and we have a short exact sequence $0 \to e^2H_1 \to H_1 \to e^1H_2 \to 0$ in which the middle term has order 4 (as an abelian group). But this is absurd as the right Ξ-modules $e^2H_c = \Xi/I_c = \Xi/(t^2+t(a-1)+(a-1)^3,a^4-1)\Xi$ and $e^2H_f = \Xi/I_f = \Xi/(a^4+a^2+1,t+a^3+a^2+a)\Xi$ are each infinite. (To see this note that for instance e^2H_f contains $E/E\cap I_f = E/(a^4+a^2+1)$ as a sub

Ξ-module). Thus there can be no such 2-knot. Therefore by Lemma 12 we must have $\pi = \Phi$. □

Corollary If the quotient of a 2-knot group π by its maximal locally-finite normal subgroup T is Φ then T is either trivial or infinite. □

In fact we believe that in this case T must be trivial, but have not yet been able to prove this.

The modules H_c and H_f are realized by the extensions of Φ by $Q(1)$ presented by

$$<a,x,y,t \mid tat^{-1} = a^2,\ txt^{-1} = y,\ tyt^{-1} = xy,\ ax = xa,\ ay = ya,$$
$$x^2 = (xy)^2 = y^2>$$

and

$$<a,x,y,t \mid tat^{-1} = a^2,\ txt^{-1} = y,\ tyt^{-1} = x,\ ax = ya,\ ay = xya,$$
$$x^2 = (xy)^2 = y^2>$$

respectively. These are in fact 3-knot groups.

The group Φ is the group of Examples 10 and 11 of Fox [Fo 1962]. By "thickening" a suitable immersed ribbon D^2 in S^3 for the stevedore's knot 6_2 (the equatorial cross-section of Example 10) we may obtain a ribbon D^3 in S^4 whose boundary is this knot. Alternatively, we may construct a ribbon D^3 in S^4 with group Φ by using the equivalent presentation $<t,u,v \mid vuv^{-1} = t,\ tut^{-1} = v>$ and the method of [H:Chapter II]. (The presentations are related by $u \to ta$, $v \to t^2at^{-1}$).

Chapter 5 THE RANK 2 CASE

The key examples of 2-knots whose groups have rank 2 abelian normal subgroups are the r-twist spins of the (p,q)-torus knots (for $p^{-1}+q^{-1}+r^{-1} \leqslant 1$). Indeed all known examples are closely related to these. Although we have not been able to show that all such 2-knot groups arise in this way, we have a number of partial results that suggest strongly that this may be so. Moreover we can characterize the groups of 2-knots which arise from branched cyclic covers of twist spins of torus knots (obvious exceptions aside) as being the 3-knot groups which are also PD_4^+-groups, have centre of rank 2, some power of a weight element being central, and such that the commutator subgroup is virtually representable onto Z. En route we characterize the groups of aspherical closed Seifert fibred 3-manifolds as being PD_3-groups with subgroups of finite index which have nontrivial centre and infinite abelianization.

Our arguments in this chapter and the next rely heavily on properties of subgroups of PD-groups and of groups with large abelian normal subgroups and small cohomological dimension. In particular we use frequently the following result, which we shall refer to as "Bieri's Theorem".

Theorem [B: Theorem 8.8] *Let G be a nonabelian group with $c.d.G = n$. Then $c.d.\zeta G \leqslant n-1$, and if ζG has rank $n-1$ then G' is free.* □

We shall also make use of special features of certain matrix groups. For instance, in this chapter we shall use the fact that $SL(2,Z)'$ is a finitely generated free normal subgroup of $GL(2,Z)$, and the quotient group is dihedral of order 24 [R: Section 6.2].

Virtually central subgroups

In all the known examples of 2-knot groups with an abelian normal subgroup of rank 2, such subgroups are free abelian, central and not contained in the commutator subgroup. In this section we shall show that the last assumption almost implies the others, and in the next section we shall provide some evidence to suggest that this last assumption may always hold.

Theorem 1 *Let π be a 2-knot group with an abelian normal subgroup A of rank 2 which is not contained in π'. Then π is a PD_4^+-group, $[\pi:C_\pi(A)] \leqslant 2$ and π' is a PD_3^+-group with nontrivial centre.*

Proof The subgroup $\pi'\cap A$ is an abelian normal subgroup of rank 1, and $\pi'/\pi'\cap A$ is finitely generated, as it has finite index in π/A. A minor extension of the argument of Lemma 9 of Chapter 4 shows that π has a torsion free abelian normal subgroup of rank 1. Therefore π is a PD_4^+-group, by Theorems 3 and 4 of Chapter 3. In particular A must itself be torsion free.

The automorphisms of A preserving the subgroup $\pi'\cap A$ form a group isomorphic to a subgroup of lower triangular 2×2 matrices with rational coefficients, which must be metabelian. Therefore π'' is contained in $C_\pi(A)$, and if $C_\pi(A)$ is solvable, π is solvable and so polycyclic [B: Theorem 9.23]. If $A \cong Z^2$ then $\pi/C_\pi(A)$ is a metabelian subgroup of $GL(2,Z)$ with finite cyclic abelianization, and so finite. The group $\pi/C_\pi(A)$ is also finite if A is not finitely generated. For otherwise $3 = c.d.A \leqslant c.d.C_\pi(A) < c.d.\pi = 4$, so $C_\pi(A)$ would be abelian by Bieri's Theorem, hence π would be polycyclic and so A finitely generated, contrary to assumption. But a finite lower triangular subgroup of $GL(2,Q)$ with cyclic abelianization must have order at most 2. Thus $[\pi:C_\pi(A)] \leqslant 2$, so π' is contained in $C_\pi(A)$ and $\pi'\cap A$ is central in π'. The subgroup H of π' generated by $\pi'\cup A$ has finite index in π and so is also a PD_4^+-group. Since A is central in this group and maps onto H/π', we have $H \cong \pi'\times Z$, and so π' is a PD_3^+-group by [B: Theorem 9.11]. □

We do not know whether A need be central in π nor whether it need be finitely generated. We shall show later that if moreover π' has a subgroup of finite index which maps onto Z then it is the fundamental group of a Seifert fibred 3-manifold. The group π is then the group of a fibred 2-knot, and the subgroup A is free abelian.

Theorem 1 applies whenever $\zeta\pi$ has rank greater than 1.

Theorem 2 *The centre of a 2-knot group* π *has rank at most* 2. *If* $\zeta\pi$ *has rank* 2, *then it is not contained in* π', *so* π *is a* PD_4^+*-group, while* $\zeta\pi' = \pi'\cap\zeta\pi$, *has rank* 1 *and is contained in* π''.

Proof We may assume that the rank of $\zeta\pi$ is greater than 1. Then π has a torsion free central subgroup of the same rank, and so is a PD_4^+-group. Now $\pi'\cap\zeta\pi$ is nontrivial, and is contained in π'', since $\pi/\pi' \cong Z$. In particular π' is nonabelian and π'' has nontrivial centre. Since π' is the fundamental group of an aspherical open 4-manifold, $c.d.\pi \leqslant 3$. In fact $c.d.\pi' = 3$, since $4 = c.d.\pi \leqslant c.d.\pi' + c.d.(\pi/\pi') \leqslant 3+1$. Now if π' is a nonabelian group with $c.d.\pi' = 3$ and $\pi''\cap\zeta\pi'$ nontrivial then $\zeta\pi'$ must have rank 1. For otherwise π'' would be free, by Bieri's Theorem, and so would be infinite cyclic and central. Hence π' would be nilpotent and of Hirsch length at most 3, by [B: Theorem 7.10]. But such a nilpotent group must be either abelian or have centre of rank 1 after all. Since $\zeta\pi'$ contains $\pi'\cap\zeta\pi$ and is characteristic in π, it follows that $\pi'\cap\zeta\pi = \zeta\pi'$ and since $\zeta\pi$ is not contained in π' it must have rank 2. □

A vacuous case?

In this section we shall suppose that the 2-knot group π has a torsion free abelian normal subgroup of rank 2 which is contained in π'. (The group π is then a PD_4^+-group). We shall also suppose that A is *maximal* i.e. is not properly contained in any other abelian normal subgroup of π. (As we shall determine all the 2-knot groups with abelian normal subgroups of rank greater than 2 in Chapter 6, this assumption is harmless). We shall first show that such a subgroup must be finitely generated. Our subsequent strategy is to attempt to show that $v.c.d.\pi/A$ is finite, and then to deduce a contradiction, in order to show that there are in fact no such 2-knot groups. We have not been completely successful, but we can show that neither π' nor $C_\pi(A)$ can be finitely generated.

A group is said to be *locally-nilpotent* if every finitely generated subgroup is nilpotent. In any group the union of all the locally-nilpotent normal subgroups is the unique maximal locally-nilpotent normal subgroup; this is called the *Hirsch-Plotkin radical* of the group [R: 12.1.3]. The Hirsch-Plotkin radical contains every abelian normal subgroup.

Theorem 3 *Let π be a 2-knot group with a maximal torsion free abelian normal subgroup A of rank 2 which is contained in π'. Then A is finitely generated.*

Proof Suppose that A is not finitely generated. Then $c.d.A = 3$, and $c.d.C_{\pi'}(A) \leqslant c.d.\pi' = 3$, so $A = C_{\pi'}(A)$ by Bieri's Theorem. If $C_\pi(A)$ had finite index in π then $A = C_{\pi'}(A)$ would have finite index in π' and so π would be virtually solvable. But then π would be virtually polycyclic [B: Theorem 9.23] and so A would be finitely generated [R: 5.4.12]. Thus we may assume that $C_\pi(A)$ has infinite index, hence $c.d.C_\pi(A) = 3$ also and so $A = C_\pi(A)$. Therefore π/A is a finitely generated subgroup of $Aut(A)$ and hence of $GL(2,Q)$. Noe if π'/A were a torsion group then it would be locally finite [K: page 105], so $h.d.\pi' = h.d.A = 2$ and hence $h.d.\pi = 3$, which would be absurd, as π is a PD_4^+-group. Therefore there is an element g of π' which is of infinite order modulo A. The subgroup B generated by g and A is solvable and $c.d.B = 3$. The Hirsch-Plotkin radical N of B contains A, so is not finitely generated, and $c.d.N = 3$ also. Therefore N is abelian, so $N = A$ and we may apply [B: Theorem 7.15]. Thus $L = H_2(A;Z)$ is the underlying abelian group of a subring $Z[m^{-1}]$ of Q, and the action of g on L is multiplication by a rational number a/b, where ab and m have the same prime divisors. (Note that the action of B on A by conjugation is determined by that of g). But g acts on A as an element of $GL(2,Q)'$, which is contained in $SL(2,Q)$, and so acts on $L = A \wedge A$ [R: 11.4.16] as $det(g) = 1$. Therefore $L \cong Z$. It follows that our supposition was false and A must be finitely generated. □

If there is such a 2-knot group then π/A is finitely presentable and maps onto Z, and so is an HNN extension with finitely generated base and associated subgroups [BS 1978]. The preimage of the base in π is a finitely generated group of cohomological dimension 3 which contains A as a normal subgroup and therefore the base has a finitely generated free subgroup of finite index [B: Theorem 8.4]. This seems a very strong condition. Nevertheless some additional information is needed in order to show that $v.c.d.\pi/A$ is finite and so to derive contradictions as in the next two theorems. (Peter M. Neumann has constructed an HNN extension with base

$\langle a,b \mid a^5 = 1\rangle$ and associated subgroups free of rank 5, and such that every subgroup of finite index contains the base. This example clearly does not have finite virtual cohomological dimension).

Lemma 1 *Let G be a solvable group with $c.d.G = 3$ and with ζG of rank at least 2. Then G is virtually abelian.*

Proof By Bieri's Theorem we may assume that $G' \cong Z$. Let $H = C_G(G')$. Then $[G:H] \leqslant |Aut(Z)| = 2$, and so $c.d.H = 3$. Moreover H is nilpotent, and the rank of ζH is at least 2, and so H must be abelian. □

Theorem 4 *Let π be a 2-knot group with a maximal abelian normal subgroup A which is free abelian of rank 2 and is contained in π'. Then π' is not finitely generated.*

Proof Suppose that π' is finitely generated. Then as $c.d.\pi' = 3$ the quotient π'/A is infinite, and has a finitely generated free subgroup of finite index [B: Theorem 8.4]. Therefore there is a subgroup B of finite index in π', which we may assume is characteristic in π' and so normal in π, such that B/A is free. There is then a subgroup H of finite index in π which contains B and such that $H/B \cong Z$. Clearly $c.d.H/A \leqslant 2$, and so H/A is a PD_2-group by [B: Theorem 9.11] (and a spectral sequence corner argument to identify the dualizing module). Since the only PD_2-groups with nontrivial finitely generated free normal subgroups are solvable, π is virtually solvable. Since $\pi/C_\pi(A)$ is isomorphic to a subgroup of $GL(2,Z)$ it has virtual cohomological dimension at most 1, and therefore $c.d.C_\pi(A) \geqslant 3$. Therefore either $C_\pi(A)$ or $C_{\pi'}(A)$ is a solvable group of cohomological dimension 3 with a central subgroup of rank 2. Therefore π has an abelian normal subgroup of cohomological dimension 3, by Lemma 1. But this contradicts the maximality of A and so our assumption that π' is finitely generated must have been wrong. □

In particular, no such group can be the group of a fibred 2-knot.

Lemma 2 *If E is a finitely generated subgroup of $GL(2,Z)$ such that $E/E' \cong Z$ then $E \cong Z$.*

Proof Every subgroup of $GL(2,Z)$ contains a free subgroup with quotient a subgroup of D_{12}, the dihedral group of order 24. If the abelianization is cyclic then this quotient must be either cyclic (of order dividing 12) or $D_3 = S_3$. But an extension of a finite cyclic group by a free group which has infinite cyclic abelianization must be torsion free, and hence free. If an extension E of D_3 by Z has infinite cyclic abelianization then it must be the group with presentation $<a,t \mid tat^{-1} = a^2, a^3 = 1>$. But then t^2 commutes with a in $SL(2,Z)$ and so must have finite order. Thus we may assume that E is an extension of D_3 by a free group $F(s)$ of rank $s > 1$. Moreover E must have cyclic subgroups of order 3, but no other finite subgroups. Since $F(s)$ has infinitely many ends so does E, and therefore $E = G*H$ or $G*_CH$ or $G*_C$, where $C = Z/3Z$, by [St: 5.A.10]. But if $E = G*H$ then either G or H would be perfect, which is clearly impossible for nontrivial subgroups of E. If $E = G*_CH$ or $G*_C$ then G (say) would have abelianization $Z/3Z$. Furthermore $G \neq Z/3Z$ and G' would be free. But then $M = G'/G''$ would be a finitely generated Z-torsion free module over the ring $Z[C] = Z[x]/(x^3-1)$ such that $M = (x-1)M$. Such a module must be 0, contradicting $G \neq Z/3Z$. Thus there is no such group, and so E must be free and therefore infinite cyclic. □

Theorem 5 *Let π be a 2-knot group with a maximal torsion free abelian normal subgroup A which is free abelian of rank 2 and is contained in π'. Then $C_\pi(A)$ has infinite index in π but is not contained in π', and is not finitely generated.*

Proof If $C_\pi(A)$ is contained in π', then $\pi/C_\pi(A)$ is isomorphic to a subgroup of $GL(2,Z)$ with infinite cyclic abelianization, and so is infinite cyclic, by Lemma 2. Therefore we may assume that either $C_\pi(A) = \pi'$ or $C_\pi(A)$ has finite index in π. In either case, $C = C_{\pi'}(A)$ has finite index in π' and so $c.d.C = 3$. Since π' is not finitely generated, by Theorem 4, π cannot be a virtually poly-Z-group. Therefore C' is a nonabelian free group, by Bieri's Theorem, and so $A \cap C' = 1$. Then A maps injectively to

$M = C/C'$. Since π/C is a finite extension of Z the ring $Z[\pi/C]$ is noetherian and M is finitely generated as a module over this ring. Moreover it has a submodule M_1 of finite index such that $M_1 = A \oplus (M_1/A)$ as an abelian group. Let P be the preimage of M_1 in π. Then $P \cong A \times (P/A)$, so $c.d.P/A = 1$ by [B: Theorem 5.6]. Since P has finite index in π' it follows that $v.c.d.\pi/A$ is finite, and hence that π/A is virtually a PD_2-group, by [B: Theorem 9.11]. Now each virtual PD_2-group with infinite abelianization maps onto some planar discontinuous group, with finite kernel [EM 1982]. In this case the planar discontinuous groups are virtually surface groups and so have compact fundamental region. On considering the presentations of such groups, as given for instance in [ZVC: Theorem 4.5.6], we see that no such group has infinite cyclic abelianization. This gives a contradiction and so proves the first assertion of the theorem.

If $C_\pi(A)$ is finitely generated and of infinite index in π, then it has cohomological dimension 3 (as above), and so $C_\pi(A)$ has a subgroup B of finite index (which we may assume is normal in π) such that B/A is a finitely generated free group. The quotient $\pi/C_\pi(A)$ is isomorphic to a subgroup of $GL(2,Z)$ and so is also finitely generated free by finite. Therefore π/B is finite-by-free-by-finite. But such a group is in fact free-by-finite, and so π/A is free-by-free-by-finite. Therefore $v.c.d.\pi/A$ is finite and π/A has a nontrivial finitely generated free normal subgroup, so we reach the same contradiction as in Theorem 4. Since $C_\pi(A)/C_{\pi'}(A)$ is cyclic it follows that $C_{\pi'}(A)$ cannot be finitely generated either. □

There are surely no 2-knot groups of the type considered in this section (even if we allow the subgroup A to have torsion).

Seifert fibred 3-manifolds

The only known examples of PD_3-groups with nontrivial centres are the fundamental groups of aspherical closed Seifert fibred 3-manifolds. Such manifolds are finitely covered by orientable circle bundles over aspherical surfaces, and so their groups have subgroups of finite index with infinite abelianization. We shall show that this extra condition is sufficient to characterize these groups. In the next section we shall use this to characterize the groups of twist spun torus knots.

Theorem 6 *Let G be a PD_3-group which has a subgroup of finite index with nontrivial centre and infinite abelianization. Then G is the fundamental group of an aspherical closed Seifert fibred 3-manifold.*

Proof A subgroup of finite index in a group with these properties has them also. On the other hand a finite torsion free extension of the group of a closed Seifert fibred 3-manifold is again the group of a closed Seifert fibred 3-manifold, by [Z: Section 63] and [Sc 1983]. Thus we may pass to subgroups of finite index whenever convenient, without loss of generality. In particular we may assume that G is orientable, G/G' is infinite and ζG is nontrivial.

Fix an epimorphism $\theta:G \to Z$ and let $K = ker(\theta)$. Suppose first that $\theta(\zeta G) = nZ$ for some nonzero n. Then $\theta^{-1}(nZ)$ is a subgroup of finite index in G which splits as a direct product $K \times Z$. Therefore K is a PD_2-group, by [B: Theorem 9.11], and so is the fundamental group of an aspherical closed surface [EM 1980], F say. Therefore $\theta^{-1}(nZ) = \pi_1(F \times S^1)$.

Now suppose that ζG is contained in K. Since G/K is infinite, $c.d.K = 2$ [St 1977]. Therefore by Bieri's Theorem either K is abelian or $\zeta G = \zeta K \cong Z$ and K' is free. If K' is infinite cyclic then G is a poly-Z group and the theorem is true. Therefore we may assume that K' is a nonabelian free group, so $\zeta G \cap K' = 1$. Since G is finitely generated, $M = K/K'$ is finitely generated as a module over $Z[G/K] = Z[t,t^{-1}]$. The image of ζG in $M \otimes Q$ generates a submodule isomorphic to $Q[t,t^{-1}]/(t-1)$. If $(t-1)^r$ is the highest power of $t-1$ dividing the order of the $Q[t,t^{-1}]$-torsion submodule of $M \otimes Q$, then the image of ζG in $N = M/((t-1)^r M, zM)$ is nontrivial, since it is nontrivial in $N \otimes Q = (M \otimes Q)/(t-1)^r(M \otimes Q)$. Since N is a finitely generated torsion free abelian group, ζG maps isomorphically to an abelian group direct summand of a subgroup N_1 of finite index in N. Let H be the inverse image of N_1 in K. Then $H \cong (H/\zeta G) \times \zeta G$ and so $H/\zeta G$ is isomorphic to a subgroup of K and has finite cohomological dimension. Let G_1 be the subgroup generated by H and some element not in K. Then G_1 has finite index in G and so is also a PD_3-group. Moreover $G_1/\zeta G$ is an HNN extension of $H/\zeta G$ and so also has finite cohomological dimension. Therefore $G_1/\zeta G$ is a PD_2-group by [B: Theorem 9.11] and so is the fundamental group of an aspherical closed surface [EM 1980], F say. Hence G_1 is the fundamental group of an S^1-bundle over F and the theorem follows. □

Extensions of this theorem characterizing the groups of closed 3-manifolds with *Sol*-structures, and of bounded Seifert fibred 3-manifolds are given in [Hi 1985]. We might expect any PD_3-group with nontrivial centre to have a subgroup of finite index with infinite abelianization, and so to be the group of a Seifert fibred 3-manifold, but we have been unable to prove this.

Twist spins of torus knots

The commutator subgroup of the group of the r-twist spin of a classical knot K is the fundamental group of the r-fold cyclic branched cover of S^3, branched over K [Ze 1965]. The r-fold cyclic branched cover of a sum of knots is the connected sum of the r-fold cyclic branched covers of the factors, and is irreducible if and only if the knot is prime. Moreover the cyclic branched covers of a prime knot are either aspherical or finitely covered by S^3; in particular no summand has free fundamental group [Pl 1984]. The cyclic branched covers of prime knots with nontrivial companions are Haken 3-manifolds [GL 1984]. The r-fold cyclic branched cover of a simple nontorus knot is a hyperbolic 3-manifold if $r \geqslant 3$, excepting only the 3-fold cyclic branched cover of the figure eight knot, which is a flat 3-manifold [Du 1983]. The r-fold cyclic branched cover of the (p,q)-torus knot is the Brieskorn Seifert fibred 3-manifold $M(p,q,r)$ which may be obtained by intersecting the unit sphere in C^3 with the complex algebraic surface $\{(u,v,w)\ in\ C^3 \mid u^p+v^q+w^r = 0\}$. (From this description it is clear that $M(p,q,r)$ is unchanged by a permutation of p, q, r). The fundamental group of $M(p,q,r)$ is finite if and only if $p^{-1}+q^{-1}+r^{-1} > 1$ [Mi 1975]. The triples $(2,q,2)$ give lens spaces, with cyclic fundamental groups. The remaining 6 such triples (with $(p,q) = 1$) lead to 3 distinct manifolds, with groups $Q(1)$, $T = T(1)$ or I^*. It follows easily from these observations and Dunbar's work that no 2-knot whose group has commutator subgroup $Q(1)\times Z/nZ$ for some $n > 1$ can be a twist spin. (Note that the meridional automorphism of such a commutator subgroup has order 6). The fundamental group of $M(2,3,6)$ is nilpotent; all the other aspherical Brieskorn 3-manifolds are finitely covered by circle bundles over surfaces of hyperbolic type, and so their fundamental groups do not have abelian normal subgroups of rank greater than 1.

If $p^{-1}+q^{-1}+r^{-1} \leqslant 1$ then the Seifert fibration of $M(p,q,r)$ is essentially unique. (Cf. Theorem 3.8 of [Sc 1983']). Let $p = ap'$, $q = bq'$ and $r = cp'q'$ where $(a,cq') = (b,cp') = (p,q) = 1$. Then $M(p,q,r)$ has b exceptional fibres of multiplicity p', a exceptional fibres of multiplicity q' and 1 exceptional fibre of multiplicity c, and so the triple $\{p, q, r\}$ is determined by the Seifert structure of $M(p,q,r)$.

Theorem 7 *Let M be the r-fold cyclic branched cover of S^3, branched over a 1-knot K, and suppose that $r > 2$ and that $\pi_1(M)$ has a subgroup of finite index with nontrivial centre and infinite abelianization but is not virtually abelian. Then K is a torus knot, and is determined uniquely by M and r.*

Proof The assumptions imply that $\pi_1(M)$ has one end, and so M is aspherical. If it were a connected sum then one of the summands would have to be a homotopy sphere, and would be a cyclic branched cover of S^3, branched over a knot summand of K. But any such homotopy 3-sphere must be standard, by the proof of the Smith conjecture. Therefore M is irreducible, and almost sufficiently large, and so must be Seifert fibred [Sc 1983]. Now since M is not a flat 3-manifold we may assume that there is a Seifert fibration which is preserved by the automorphisms of the branched cover [MS 1986]. Since $r > 2$ the fixed circle (the branch set in M) must be a fibre of the fibration, which therefore passes to a Seifert fibration of $X(K)$. Thus K must be a torus knot [BZ 1967], and so M is a Brieskorn manifold. The uniqueness follows as in the above paragraph. □

If $\pi_1(M)$ is virtually abelian and infinite then a similar argument shows that M is irreducible, hence flat, and that only $r = 3$ or 4 is possible. It is not hard to show that only one of the six closed orientable flat

3-manifolds is a cyclic branched cover of S^3, branched over a knot, and the order of the cover must be 3. (See Theorems 6 and 8 of Chapter 6). Theorems 2 and 7 together with this observation imply that a classical knot k such that $\pi\tau_r k$ has centre of rank 2 and whose commutator subgroup has a subgroup of finite index which maps onto Z (for some $r \geqslant 3$) must be a torus knot.

It is in fact clear from the tables of [Du 1983] that if the r-fold cyclic branched cover of a classical knot is Seifert fibred for some $r \geqslant 3$ then either the knot is a torus knot or it is the figure eight knot and $r = 3$. All the knots whose 2-fold branched covers are Seifert fibred are torus knots or Montesinos knots. (This class was first described in [Mo 1973], and includes the 2-bridge knots and pretzel knots). The number of distinct knots whose 2-fold branched cover is a given Seifert fibred 3-manifold can be arbitrarily large [Be 1984]. Moreover for each $r \geqslant 2$ there are distinct simple 1-knots whose r-fold cyclic branched covers are homeomorphic [Sa 1981, Ko 1986].

Pao has observed that if K is a fibred 2-knot with monodromy of finite order r and if $(r,s) = 1$ then the s-fold cyclic branched cover of S^4, branched over K, is again a 4-sphere and so the branch set gives a new 2-knot, which we shall call the s-fold cyclic branched cover of K. This new knot is again fibred, with the same fibre and monodromy the sth power of that of K [Pa 1978, Pl 1983]. Using properties of S^1-actions on smooth homotopy 4-spheres, Plotnick obtains the following result.

Theorem [Pl 1986] *Modulo the 3-dimensional Poincaré conjecture, the class of all fibred 2-knots with periodic monodromy is precisely the class of all s-fold cyclic branched covers of r-twist spins, where $0 < s < r$ and $(r,s) = 1$.* □

Here "periodic monodromy" means that the fibration of the exterior of the knot has a characteristic map which is of finite order. It is not in general sufficient that the *closed* monodromy be represented by a map of finite order. We shall call such knots *branched twist spins*, for short.

If we restate this result in terms of twist spinning knots in homotopy 3-spheres we may avoid explicit dependence on the Poincaré conjecture. In our application in the next theorem we are able to show directly that the homotopy 3-sphere arising there may be assumed to be standard.

Theorem 8 *A group G which is not virtually solvable is the group of a 2-knot which is a cyclic branched cover of a twist spun torus knot if and only if it is a 3-knot group, a* PD_4^+*-group with centre of rank* 2, *some nonzero power of a weight element being central, and* G' *has a subgroup of finite index with infinite abelianization.*

Proof If K is a cyclic branched cover of the r-twist spin of the (p,q)-torus knot then $M(K)$ fibres over the circle with fibre $M(p,q,r)$ and monodromy of order r, and so the rth power of a meridian is central. Moreover the monodromy commutes with the natural circle action on $M(p,q,r)$ (cf. Lemma 1.1 of [Mi 1975]) and hence preserves a Seifert fibration. It follows that the meridian commutes with the centre of $\pi_1(M(p,q,r))$, which is therefore also central in G. Since (with the above exceptions) $\pi_1(M(p,q,r))$ is a PD_3^+-group with infinite cyclic centre and which is virtually representable onto Z, the necessity of the conditions is evident.

Conversely, if G is such a group then G' is the fundamental group of a Seifert fibred 3-manifold, N say, by Theorems 2 and 6. Moreover N is "sufficiently complicated" in the sense of [Zi 1979], since G' is not virtually solvable. Let t be an element of G whose normal closure is the whole group, and such that t^n is central for some positive n (which we may assume minimal). Then $\hat{G} = G/\langle t^n \rangle$ is a semidirect product of Z/nZ with the normal subgroup G'. By Corollary 3.3 of [Zi 1979] there is a fibre preserving selfhomeomorphism τ of N inducing the outer automorphism of G' determined by t, which moreover can be so chosen that its lifts to the universal covering space $\tilde{N}$ together with the covering group generate a group of homeomorphisms isomorphic to $\hat{G}$. The automorphism of $\tilde{N}$ corresponding to the image of t in $\hat{G}$ has a connected 1-dimensional fixed point set by Smith theory. (Note that $\tilde{N} = R^3$). Therefore the fixed point set of the map τ in N is nonempty. Let P be a fixed point. Then P determines a crosssection $\gamma = P\times S^1$ of the mapping torus of τ. We may perform surgery on γ to obtain a 2-knot with group G which is fibred with monodromy (of the fibration for the exterior X) of finite order. We may then apply the above theorem of Plotnick to conclude that the 2-knot is a branched twist spin of a knot in a homotopy 3-sphere. Since the monodromy respects the Seifert fibration and leaves the centre of G' invariant, the branch set must be a fibre, and the orbit manifold a Seifert fibred homotopy 3-sphere. Therefore the orbit knot is a fibre of a Seifert fibration of S^3 and so is a torus knot. Thus the 2-knot is a branched twist spin of a torus knot. □

It shall follow from the work of Chapter 6 that if π is a virtually solvable 2-knot group in which some power of a weight element is central then either π' is finite or π is the group of the 2-twist spin of a Montesinos knot $K(0\mid b;(3,1),(3,1),(3,\pm1))$ (with b even) or of the 3-twist spin of the figure eight knot or of the 6-twist spin of the trefoil knot.

We might hope to avoid the appeal to S^1-actions implicit in our use of Plotnick's theorem by a homological argument to show that τ has connected fixed point set. The projection onto the orbit space would then be a cyclic branched covering, branched over a knot, and we might then be able to show that the orbit space is simply connected, by using the fact that the normal closure of t in G is the whole group. Since N is Seifert fibred and τ is fibre preserving, the branch locus would then be a torus knot in the standard 3-sphere. However we have been unable thus far to establish this connectedness directly.

If p, q and r are relatively prime then $M(p,q,r)$ is an homology sphere and the group π of the r-twist spin of the (p,q)-torus knot has a central element which maps to a generator of π/π'. It follows that $\pi \cong \pi' \times Z$, and π' has weight 1. Moreover if t is a generator for the Z summand, then an element h of π' is a weight element for π' if and only if ht is a weight element for π. This correspondance also gives a bijection between conjugacy classes of such weight elements. If we exclude the case (2,3,5) then π' has infinitely many distinct weight orbits, and moreover there are weight elements such that no power is central [Pl 1983]. Therefore we may obtain many 2-knots whose groups are as in Theorem 8 but which are not themselves branched twist spins, by surgery on such weight classes in $M(p,q,r)\times S^1$. Is there a 2-knot group which contains a rank 2 free abelian normal subgroup and for which no nonzero power of any weight element is central?

If K is a 2-knot with group as in Theorem 8 then $M(K)$ is aspherical, and so the theorem implies that it is homotopy equivalent to $M(K_1)$ for some K_1 which is a branched twist spin of a torus knot. It is a well known open question as to whether homotopy equivalent aspherical closed manifolds must be homeomorphic. In Chapter 8 we shall see that $M(K)$ and $M(K_1)$ must be topologically s-cobordant. Here we shall show that if K is fibred, with irreducible fibre, then $M(K)$ and $M(K_1)$ are homeomorphic. This is a version of Proposition 6.1 of [Pl 1986], starting from more algebraic hypotheses.

Theorem 9 *Let K be a fibred 2-knot whose group π has centre of rank 2, some power of a weight element being central, and such that π' has a subgroup of finite index with infinite abelianization. Suppose that the fibre is irreducible. Then $M(K)$ is homeomorphic to $M(K_1)$ where K_1 is some branched twist spin of a torus knot.*

Proof Let F be the closed fibre and $\phi:F\rightarrow F$ the characteristic map. Then F is a Seifert fibred manifold, as above. Now the automorphism of F constructed as in Theorem 8 induces the same outer automorphism of $\pi_1(F)$ as ϕ, and so these maps must be homotopic. Therefore they are in fact isotopic, by [Sc 1985, BO 1986]. The theorem now follows. □

The closed fibre of any fibred 2-knot with such a group is aspherical, and so is a connected sum $F\#P$ where F is irreducible and P is a homotopy 3-sphere. If we could show that the characteristic map may be isotoped so that the fake 3-cell (if there is one) is carried onto itself then we would have $K = K_1\#K_2$ where K_1 is fibred with irreducible fibre and K_2 has group Z. We could then use Freedman's Unknotting Theorem to sidestep the assumption that the fibre be irreducible.

Other twist spins

We may also apply Plotnick's theorem in attempting to understand twist spins of other knots. As the arguments are similar to those of Theorems 8 and 9, except in that the existence of homeomorphisms of finite order and "homotopy implies isotopy" require different justification, while the conclusions are less satifactory, we shall not give proofs for the following assertions. (Cf. also Theorem 9 of Chapter 2).

Let G be a 3-knot group which is a PD_4^+-group in which some nonzero power of a weight element is central. If G' is the fundamental group of a hyperbolic 3-manifold and the 3-dimensional Poincaré conjecture is true then we may use the Rigidity Theorem of Mostow [Mo 1968] to show that G is the group of some branched twist spin K of a simple non-torus knot. Moreover if K_1 is any other fibred 2-knot with group G and hyperbolic fibre then $M(K_1)$ is homeomorphic to $M(K)$. In particular the simple knot and the order of the twist are uniquely determined.

Similarly if G' is the fundamental group of a Haken, non-Seifert fibred 3-manifold and the 3-dimensional Poincaré conjecture is true then we may use [Zi 1982] to show that G is the group of some branched twist spin of a prime nontorus knot. If moreover all finite group actions on the fibre are geometric then by [Zi 1986] the prime knot and the order of the twist are unique. However we do not yet have algebraic characterizations of the groups of hyperbolic or Haken 3-manifolds comparable to Theorem 6. (Question 18 of [Th 1982] asks whether every closed hyperbolic 3-manifold is finitely covered by a surface bundle over the circle).

We raise the following questions. Is a 3-knot group which is a PD_4^+-group in which some nonzero power of a weight element is central the group of a branched twist spin of a prime knot? If K is a 2-knot such that π' has one end and some power of a weight element is central in π then is $M(K)$ homeomorphic to a cyclic branched cover of $M(\tau_r k)$ for some prime classical knot k and $r \geqslant 2$, and if so are k and r unique?

Chapter 6 ASCENDING SERIES AND THE LARGE RANK CASES

All but two of the 2-knot groups with abelian normal subgroups of rank greater than 2 have commutator subgroup Z^3 (as in the examples of Cappell and Shaneson); the exceptions are virtually abelian of rank 4. We shall prove this after considering the more general class of groups with ascending series whose factors are either locally-finite or locally-nilpotent. If π is such a 2-knot group and T is its maximal locally-finite normal subgroup then π/T is in fact either Z or Φ or virtually poly-Z of Hirsch length 4. In the latter case the Hirsch-Plotkin radical of π'/T must be either Z^3 or nilpotent of class 2, and has finite index in π'/T. Moreover the natural homomorphism from the quotient to the outer automorphism group of the Hirsch-Plotkin radical is injective.

When the Hirsch-Plotkin radical is Z^3 the image of this homomorphism is either trivial, the four-group V or the alternating group A_4. If moreover π is torsion free then π' is either Z^3 or G_6, the group of the orientable flat 3-manifold with noncyclic holonomy. These conditions hold if π has an abelian normal subgroup of rank greater than 2. We then complete the classification of such 2-knot groups by determining the possible meridianal automorphisms of π'.

When the Hirsch-Plotkin radical is nilpotent of class 2 the image of this homomorphism is either trivial, $Z/3Z$ or V. If π' is virtually nilpotent then it must be either torsion free or finite, and we describe all such groups. All the virtually solvable 2-knot groups with abelian normal subgroups of rank greater than 0 are of this form. (We believe that these comprise *all* the 2-knot groups with ascending series as above, for it seems likely that if π' is infinite then T is trivial).

Generalisations of solvability

Freedman has shown that the class of groups for which TOP surgery techniques work in dimension 4 includes all finite groups and Z and is closed under the operations of extension, taking sub- and quotient groups, and increasing countable union. Let F be the smallest such class. As the property of having no noncyclic free subgroups is preserved under these operations, no group in F has noncyclic free subgroups, and so all finitely generated groups in F have finitely many ends. Other properties shared by members of F are hard to find. This is in part because a normal subgroup of one group need not even be subnormal in a larger group.

We may obtain a class which is more tractable in this respect by considering the smallest subclass G of F which contains all locally-finite groups and Z and is closed under extension and increasing countable *subnormal* union (i.e. if G_n is in G and G_n is a normal subgroup of G_{n+1} for all n in N then $\cup G_n$ is in G). It is not hard to see that the property of having either a nontrivial locally-finite normal subgroup or a nontrivial Hirsch-Plotkin radical is preserved under these operations. In fact G may also be described as the class of groups which have ascending series [R: page 344] with factors either locally-finite or locally-nilpotent. This class clearly contains all countable groups which are extensions of solvable groups by locally-finite normal subgroups. Although it certainly has other members, we shall show that all 2-knot groups in G are such extensions.

Theorem 1 *Let π be a 2-knot group which has an ascending series with factors which are either locally-finite or locally-nilpotent, and let T be the maximal locally-finite normal subgroup of π. Then either $\pi/T \cong Z$ or Φ or π/T is a PD_4^+-group over Q.*

Proof Clearly T is contained in π' and so $J = \pi/T$ is nontrivial, and $J/J' \cong Z$. We may assume that $J \neq Z$. The ascending series for π induces a similar series for J; by the maximality of T the Hirsch-Plotkin radical B of J' must be nontrivial, and moreover it must be torsion free, as the torsion subset of a locally-nilpotent group is a characteristic locally-finite subgroup [R: 5.2.7]. Therefore it has a nontrivial maximal torsion free abelian normal subgroup A, and the result follows from Theorem 6 of Chapter 3 except perhaps when A has rank 1 and $e(J/A) = 1$. It then follows that A is central in J' and that J/A is infinite and has no nontrivial locally-finite normal subgroup. For the preimage P of such a subgroup in J would be a central extension of a locally-finite group and so would have locally-finite commutator subgroup (by an easy extension of Schur's theorem [R: 10.1.4]). As the preimage of P' in π is then a locally-finite subgroup, P' must be trivial (by maximality of T) and so $P = A$ (by maximality of A). Therefore J/A must have instead a nontrivial Hirsch-Plotkin radical. In particular there is a characteristic subgroup C of B which contains A

properly (and so is nonabelian), and for which C/A is abelian. Choose two noncommuting elements x, y of C and let D be the subgroup that they generate. Then D is poly-Z of Hirsch length 3, and $Q[J]$ is a free $Q[D]$-module, so $H^s(D;Q[J]) = 0$ for $s \leqslant 2$. applying the LHS spectral sequence for the extensions of DA/D by D, C/DA by DA and J/C by C in turn, we conclude that $H^s(J;Q[J]) = 0$ for $s \leqslant 2$. Therefore J is a PD_4^+-group over Q by the Corollary to Theorem 3 of Chapter 3. □

We showed in Theorem 6 of Chapter 3 that if $\pi/T \cong Z$ then T must be finite. We expect that in fact in all other cases T must be trivial. Note that it follows from our proof that if π/T is a PD_4^+-group then the Hirsch length of the Hirsch-Plotkin radical of π'/T is at least 2. In the next section we shall see that π/T is in fact virtually poly-Z.

The Hirsch-Plotkin radical

We consider next which PD_4^+-groups over Q can arise as such quotients π/T of 2-knot groups with such ascending series. For each natural number q let Γ_q be the group with presentation $\langle x,y,z \mid [x,y] = z^q,\ xz = zx,\ yz = zy\rangle$. Then Γ_q is a torsion free 2-stage nilpotent group with centre $\zeta\Gamma_q = Z$ (generated by the image of z) and $\Gamma_q/\zeta\Gamma_q = Z^2$ (generated by the images of x and y). Every finitely generated torsion free nilpotent group of Hirsch length 3 is isomorphic to Z^3 or to Γ_q for some q.

Lemma 1 *Let D be a torsion subgroup of $GL(2,Q)$. Then D is finite, of order at most* 12.

Proof Let E be a finite subgroup of D. If L is a lattice in Q^2 then $\cup e(L)$ is a lattice invariant under E, so E is conjugate to a (finite) subgroup of $GL(2,Z)$ and hence has order at most 12 (cf. [Z: page 85]). As D is locally finite [K: page 105], it is an increasing union of its finite subgroups and so also has order at most 12. □

Theorem 2 *Let J be a PD_4^+-group over Q with $J/J' \cong Z$ which has* *an ascending series with factors either locally-finite or locally-nilpotent, and suppose that J has no nontrivial locally-finite normal subgroup. Then J' is a finite extension of Z^3 or of Γ_q (for some natural number q).*

Proof Let B be the Hirsch-Plotkin radical of J', and let h be the Hirsch length of B. As in Theorem 1, B is torsion free and h is at least 2. Since $[J:J']$ is infinite, $h.d._Q J' < h.d._Q J = 4$ by [B: Theorem 9.22], and so $h \leqslant h.d._Q B \leqslant h.d._Q J' \leqslant 3$. Suppose that $h = 2$. Then B is locally-abelian, hence abelian, of rank 2. Let E be the preimage in J of the maximal locally-finite normal subgroup of J/B and let $\alpha:E \rightarrow Aut(B)$ be the homomorphism induced by conjugation. Then B is central in $ker(\alpha)$, and Schur's Theorem then implies that $B = ker(\alpha)$ (as in Theorem 1). Moreover $im(\alpha)$ is a locally finite subgroup of $Aut(B)$, which is in turn isomorphic to a subgroup of $GL(2,Q)$. Therefore by Lemma 1 $im(\alpha)$ is finite and so E/B is a finite solvable group. Now $J' \neq E$, for otherwise $h.d._Q J \leqslant h.d._Q B + 1$ (assuming J'/B is locally-finite) $\leqslant 3$ which is absurd, as J is a PD_4^+-group over Q. Let F be the preimage in J' of the maximal abelian normal subgroup of J'/E. Then F/E is nontrivial and torsion free. Moreover F is solvable and $h.d._Q F \leqslant h.d._Q J' \leqslant 3$ and so the Hirsch length of F is at most 3. Since it is at least $h+1$, it must equal 3 and so equal $h.d._Q F$. Moreover J'/F must be a torsion group. This has two contradictory consequences. On the one hand F/E has rank 1 and so is central in J'/E. Using Schur's Theorem and the maximality of E/B we find that $F = J'$. On the other hand $[J:F] = \infty$ so $c.d._Q F \leqslant 3$ by [St 1977]. Since F is solvable of Hirsch length 3 and virtually torsion free, this implies that F must be finitely generated [GS 1981]. But then $J'/J''E = F/F' \cong Z \oplus z(F/F')$, which is impossible for a group (J/E) with infinite cyclic abelianization. Therefore $h = 3$.

Since $c.d._Q B \leqslant 3$ by [St 1977], $h = c.d._Q B$ and so B is finitely generated [B: Theorem 7.14], and therefore must be isomorphic to Z^3 or to Γ_q for some Q. Finally let j be an element of J which is not in J'. Then the subgroup generated by $B \cup \{j\}$ is a poly-Z group of Hirsch length 4 and so has finite index in J by [B: Theorem 9.22]. Therefore J is a poly-Z group and B has finite index in J'. □

If we knew *a priori* that J was virtually poly-Z, we could simplify the proof of this theorem by observing that every sub- and quotient group is then also virtually poly-Z.

Abelian HP radical

In this section we shall show that if the Hirsch-Plotkin radical H of the commutator subgroup $J' = \pi/T$ of the quotient of a 2-knot group π by its maximal locally-finite normal subgroup T is Z^3 then J'/H is isomorphic to a finite subgroup of $SL(2,Z)$. We shall then show that J'/H must be trivial, the four-group V or the tetrahedral group A_4. In particular π/T is solvable (of derived length at most 4). If moreover π/T is torsion free then $\pi' \cong Z^3$, with just two exceptions.

As we have found no convenient reference listing the finite subgroups of $SL(3,Z)$ we shall derive what we need in our next lemma. Since any integral matrix of finite order is diagonalizable over the complex numbers the corresponding cyclotomic polynomial must divide the characteristic polynomial of the matrix. Thus in particular the finite cyclic subgroups of $SL(2,Z)$ or $SL(3,Z)$ have order 1, 2, 3, 4 or 6. In fact the only other finite subgroups of $GL(2,Z)$ are the dihedral groups of order 2, 4, 6, 8 or 12 [Z: page 85]. (Note that $-I$ is the only element of order 2 and determinant 1, but that there are two conjugacy classes of elements of order 2 and determinant -1).

Lemma 2 *Let E be a finite subgroup of $SL(3,Z)$. Then the order of E divides 24, and E is either cyclic, dihedral, a semidirect product of $Z/8Z$ or D_8 with normal subgroup $Z/3Z$, or is A_4 or S_4.*

Proof If p is an odd prime, A is a 3×3 integral matrix and $k = p^v q$ with $(p,q) = 1$ then $(I+p^r A)^k \equiv I+kp^r A$ modulo (p^{2r+v}). It follows that the kernel of the natural map from $SL(3,Z)$ to $SL(3,Z/pZ)$ is torsion free. Therefore the order of E must divide the order of $SL(3,Z/pZ)$, which is $p^3(p^3-1)(p^2-1)$, for each odd prime p, and so divides $48 = 2^4 3$.

Now suppose that there is a central element α of order 2. Since α has determinant 1 its eigenvalues must be 1, −1, −1 and we can find a basis $\{u, v, w\}$ for Z^3 such that $\alpha(u) = u$. If β is any other element of

E then $\alpha\beta(u) = \beta\alpha(u)$ (since α is central) $= \beta(u)$, so $\beta(u) = \pm u$ (since u generates $ker(\alpha-1)$ and $det(\beta) = 1$). Therefore β induces an automorphism of $Z^3/<u> \cong Z^2$, and so there is a homomorphism from E to $GL(2,Z)$ which is easily seen to be injective. By the remarks above, E must be cyclic or dihedral, of order dividing 24.

Since every nontrivial 2-group has nontrivial centre, it follows that in general any 2-subgroup of E has order dividing 8, and so the order of E again divides 24. The lemma is easily verified if E has a normal Sylow 3-subgroup. Otherwise E has order 12 or 24, and by Sylow's theorem [R: 1.6.16] it has 4 distinct subgroups of order 3, which are permuted by conjugation. The kernel of the induced homomorphism from E to S_4 is a normal subgroup of index divisible by 4. It cannot have order 3 or 6, for then E would have a normal subgroup of order 3. Nor can it have order 2, for then it would be central, so E would be cyclic or dihedral, and thus again have a normal subgroup of order 3. Therefore this homomorphism is injective. It follows easily that E must be A_4 or S_4. □

In fact not all the groups allowed by this lemma are subgroups of $SL(3,Z)$. We shall not prove this as we shall in any case impose a stringent further condition. We are only interested in those subgroups of $SL(3,Z)$ which can occur as the commutator subgroup of a group (J/H) with infinite cyclic abelianization.

Theorem 3 *Let J be a PD_4^+-group over Q with $J/J' \cong Z$ and which has no nontrivial torsion normal subgroup. If the Hirsch-Plotkin radical H of J' is isomorphic to Z^3 then either $J' = H$ or $J'/H \cong V$ or A_4.*

Proof Let j be an element of J which is not in J'. Since H is normal in J the subgroup generated by H and j is a poly-Z group, of Hirsch length 4, and so has homological dimension 4. Therefore it has finite index in J [B: Theorem 9.22] and so J'/H is finite.

The conjugation action of J on H induces a homomorphism α from J' to $Aut(H) \cong GL(3,Z)$. Since J is of orientable type as a PD_4-group over Q the image of α lies in $SL(3,Z)$ [B: page 177]. Since H is central in $ker(\alpha)$, and of finite index there, $(ker(\alpha))'$ is finite, by Schur's

Theorem [R: 10.1.4]. As $(ker(\alpha))'$ is a normal subgroup of J it must therefore be trivial, and so $ker(\alpha) = H$ (by maximality of H). Thus J'/H is isomorphic to a finite subgroup of $SL(3,Z)$, and so must be one of the groups allowed by Lemma 2. In particular, J must be solvable, and so has weight 1. Therefore J'/H must admit a meridianal automorphism and so can only be trivial, $Z/3Z$, V or A_4.

Any element of $SL(3,Z)$ of order 3 has eigenvalues 1, ω, ω^2 (where ω is a primitive cube root of unity). Therefore it is conjugate to a matrix of the form $\begin{pmatrix} 1 & (a,b) \\ 0 & \Omega \end{pmatrix}$ where $\Omega = \begin{pmatrix} 0 & -1 \\ 1 & -1 \end{pmatrix}$ is an element of order 3 in $SL(2,Z)$, for some a, b in Z. Thus if $J'/H = Z/3Z$ the group J' must have a presentation of the form

$$<H,t \mid txt^{-1} = x,\ tyt^{-1} = x^a z,\ tzt^{-1} = x^b y^{-1} z^{-1},\ t^3 = x^c y^d z^e>$$

for some a, b, c, d, e in Z. But the abelianization of such a group is $Z\oplus(torsion)$, and so admits no meridianal automorphism. Thus J'/H cannot be cyclic of order 3, and so must be 1, V or A_4. □

If such a group J is torsion free then J' is a PD_3^+-group which is virtually abelian and so is the fundamental group of an orientable flat 3-manifold. Of the six such groups (listed on page 117 of [Wo]) only Z^3 and G_6, the group with presentation $<x,y \mid xy^2x^{-1}y^2 = yx^2y^{-1}x^2 = 1>$, admit meridianal automorphisms. (For the other groups have abelianizations of the form $Z\oplus(finite)$ or $Z^r\oplus(Z/2Z)$). Our next two theorems shall give a direct proof of this without reference to the notion of flat manifold.

Theorem 4 *Let J be a group with $J/J' \cong Z$ and which has no non-trivial torsion normal subgroup. If the Hirsch-Plotkin radical H of J' is isomorphic to Z^3, and the quotient J'/H is isomorphic to V and acts orientably on H, then either $J' \cong G_6$ or J' is the semidirect product of V with Z^3 with presentation*

$$<Z^3,s,v \mid sx = xs,\ sys^{-1} = xy^{-1},\ szs^{-1} = xz^{-1},\ vxv^{-1} = x^{-1},$$

$$vyv^{-1} = x^{-1}y,\ vzv^{-1} = z^{-1},\ s^2 = v^2 = (sv)^2 = 1>.$$

Proof If A is an element of order 2 in $SL(3,Z)$, then it has eigenvalues 1, -1, -1 and so we may assume that it has the form $\begin{pmatrix} 1 & \alpha \\ 0 & M \end{pmatrix}$ for some α in Z^2 and M in $SL(2,Z)$. Moreover M has order 2 and so $M = -I$. If B is another element of order 2 which commute with A then $B = \begin{pmatrix} \varepsilon & \beta \\ 0 & N \end{pmatrix}$ for some $\varepsilon = \pm 1$ and β in Z^2 and N of order 2 in $GL(2,Z)$ with $det(N) = \varepsilon$. If $\varepsilon = 1$ then $N = -I$ and so $AB = BA$ implies that $B = A$. We may therefore assume that $\varepsilon = -1$ and hence (after conjugation in $GL(2,Z)$) that $N = \begin{pmatrix} 1 & 0 \\ 0 & -1 \end{pmatrix}$ or $\begin{pmatrix} 1 & -1 \\ 0 & -1 \end{pmatrix}$. We may reduce the entries of α modulo (2) by conjugation by elements of the form $\begin{pmatrix} 1 & \theta \\ 0 & I \end{pmatrix}$, which does not change N. Thus there are 4 possibilities for α, namely (0,0), (1,0), (0,1) or (1,1). Moreover $AB = BA$ implies that $2\beta+\alpha(N+I) = 0$. Thus up to conjugation by matrices of the form $\begin{pmatrix} 1 & \gamma \\ 0 & C \end{pmatrix}$ in $GL(3,Z)$ there are 6 pairs (A,B) that we need consider. After conjugation by permutation matrices (etc.) there remain 4 (presumably) distinct conjugacy classes (in $GL(3,Z)$) of subgroups of $SL(3,Z)$ which are isomorphic to V.

The first class is represented by the group of diagonal matrices. Any corresponding extension J' of V by $H \cong Z^3$ must have a presentation of the form

$$<Z^3,t,u \mid tx = xt,\ tyt^{-1} = y^{-1},\ tzt^{-1} = z^{-1},\ uxu^{-1} = x^{-1},\ uy = yu,$$

$$uzu^{-1} = z^{-1},\ t^2 = x^a y^b z^c,\ u^2 = x^d y^e z^f,\ (tu)^2 = x^g y^h z^i>$$

for some $a, \cdots i$ in Z. On replacing t by tx^{-n} the exponent a becomes $a-2n$ (since $tx = xt$); thus we may assume that $a = 0$ or 1, and similarly $e = 0$ or 1 and $i = 0$ or 1. Since $x^a y^b z^c = t^2 = tx^a y^b z^c t^{-1} = x^a y^{-b} z^{-c}$ we must have $b = c = 0$, and similarly $d = f = g = h = 0$. If such a group is torsion free we must then have $a = e = i = 1$. The group is then isomorphic to G_6.

If J' is not torsion free then (without loss of generality) we may assume that $t^2 = 1$. Suppose that J has a meridianal automorphism ϕ. Then ϕ induces a meridianal automorphism on $J'/H = V$. If ϕ is meridianal then so is ϕ^{-1}, and so we may assume that $\phi(t) = uh$ and $\phi(u) = tuh'$ for some h and h' in H. Then u^2 and $(tu)^2$ are also squares in H, and so

J'/J'' is an elementary 2-group with basis the images of x, y, z, t and u. Since $tx = xt$ we must have $u\phi(x) = \phi(x)u$ and so $\phi(x)$ is a power of y. Similarly $\phi(y)$ is a power of z and $\phi(z)$ is a power of x. Since ϕ is an automorphism we then have $\phi(x) \equiv y$ modulo J'' etc. But such an automorphism of $J'/J'' = (Z/2Z)^5$ cannot be meridianal. Therefore every such extension is torsion free.

The second class is represented by the group generated by the matrices

$$\begin{pmatrix}1 & 1 & 1\\ 0 & -1 & 0\\ 0 & 0 & -1\end{pmatrix} \text{ and } \begin{pmatrix}-1 & -1 & 0\\ 0 & 1 & 0\\ 0 & 0 & -1\end{pmatrix}.$$

Any corresponding extension J' of V by H must have a presentation of the form

$$<Z^3,t,u \mid tx = xt,\ tyt^{-1} = xy^{-1},\ tzt^{-1} = xz^{-1},\ uxu^{-1} = x^{-1},\ uyu^{-1} = x^{-1}y,$$

$$uzu^{-1} = z^{-1},\ t^2 = x^a y^b z^c,\ u^2 = x^d y^e z^f,\ (tu)^2 = x^g y^h z^i>$$

for some $a, \cdots i$ in Z. As before $b = c = f = h = 0$, and $2d+e = 2g+i = 0$ (and we may assume that $a = 0$ or 1). Then $(uy^d)^2 = 1$ and $(tz^{-g}uy^d)^2 = 1$, so on replacing t by $s = tz^{-g}$ and u by $v = uy^d$ we obtain a new presentation

$$<Z^3,s,v \mid sx = xs,\ sys^{-1} = xy^{-1},\ szs^{-1} = xz^{-1},\ vxv^{-1} = x^{-1},$$

$$vyv^{-1} = x^{-1}y,\ vzv^{-1} = z^{-1},\ s^2 = x^a,\ (sv)^2 = v^2 = 1>$$

for some $a = 0$ or 1. Thus J' has torsion. If ϕ is a meridianal automorphism then as before $a = 0$ and so $J'/J'' = (Z/2Z)^4$, generated by the images of y, z, s and v. We may assume that $\phi(s) = vh$ and $\phi(v) = svh'$ for some h, h' in H. Since $sx = xs$ we must have $v\phi(x) = \phi(x)v$ and so $\phi(x) = (x^{-1}y^2)^m$ for some m in Z. Similarly we find that $\phi(y) = x^n y^m z^{-m-2n}$ and $\phi(z) = x^p y^m$ for some n, p in Z. Since $\phi \mid H$ must be an automorphism, and must be orientation preserving, as every subgroup of finite index in J is also a PD_4^+-group over Q, we must have $det(\phi \mid H) = 1$. There are four possibilities for the triple (m,n,p), namely $(1,0,0)$, $(1,-1,-1)$, $(-1,0,1)$ and $(-1,1,0)$.

The third class is represented by the group generated by the matrices

$$\begin{pmatrix}1 & 0 & 0\\ 0 & -1 & 0\\ 0 & 0 & -1\end{pmatrix} \text{ and } \begin{pmatrix}-1 & 0 & 0\\ 0 & 1 & -1\\ 0 & 0 & -1\end{pmatrix}.$$

Any corresponding extension J' of V by H must have a presentation of the form

$$<Z^3,t,u \mid tx = xt,\ tyt^{-1} = y^{-1},\ tzt^{-1} = z^{-1},\ uxu^{-1} = x^{-1},\ uy = yu,$$
$$uzu^{-1} = y^{-1}z^{-1},\ t^2 = x^a y^b z^c,\ u^2 = x^d y^e z^f,\ (tu)^2 = x^g y^h z^i>$$

for some $a, \cdots i$ in Z. As before we find that we may assume that $a = 0$ or 1, $e = 0$ or 1 and $b = c = d = f = g = 0$. We also obtain $i = 2h$. Therefore $(tz^h u)^2 = 1$. On replacing t by $s = tz^h$, we obtain a new presentation

$$<Z^3,s,u \mid sx = xs,\ sys^{-1} = y^{-1},\ szs^{-1} = z^{-1},\ uxu^{-1} = x^{-1},\ uy = yu,$$
$$uzu^{-1} = y^{-1}z^{-1},\ u^2 = y^e,\ s^2 = x^a,\ (su)^2 = 1>$$

for some $a = 0$ or 1 and $e = 0$ or 1. Thus J' must have torsion. If it admits a meridianal automorphism ϕ then we may assume that $\phi(s) = uh$ and $\phi(u) = suh'$ for some h, h' in H. But $\phi(s)\phi(x) = \phi(x)\phi(s)$ then implies that $\phi(x) = y^n$ for some n in Z. Since ϕ is an automorphism $n = \pm 1$. But x is in J'' whereas y is not, so we have a contradiction. Thus there are no groups J corresponding to this third class.

The final class is represented by the group generated by the matrices

$$\begin{pmatrix}1 & 0 & 1\\ 0 & -1 & 0\\ 0 & 0 & -1\end{pmatrix} \text{ and } \begin{pmatrix}-1 & 0 & 0\\ 0 & 1 & -1\\ 0 & 0 & -1\end{pmatrix}.$$

Any corresponding extension J' of V by H must have a presentation of the form

$$<Z^3,t,u \mid tx = xt,\ tyt^{-1} = y^{-1},\ tzt^{-1} = xz^{-1},\ uxu^{-1} = x^{-1},\ uy = yu,$$
$$uzu^{-1} = y^{-1}z^{-1},\ t^2 = x^a y^b z^c,\ y^2 = x^d y^e z^f,\ (tu)^2 = x^g y^h z^i>.$$

Once again we may assume that $a = 0$ or 1, $e = 0$ or 1, $b = c = d = f = 0$ and $2g = -2h = -i$, and then $(tuz^g)^2 = 1$. We then find that $J'/J'' = (Z/4Z) \oplus (Z/2Z)$ which admits no meridianal automorphisms. Therefore this case does not arise at all either. □

We shall not treat all the cases when J'/H is isomorphic to A_4.

Theorem 5 *Let J be a torsion free group with $J/J' \cong Z$. If the Hirsch–Plotkin radical H of J' is isomorphic to Z^3, and the quotient J'/H is finite and acts orientably on H then either $J' = H \cong Z^3$ or $J' \cong G_6$.*

Proof By Theorems 3 and 4 it shall suffice to consider the possibility that $J'/H \cong A_4$. The group A_4 is a semidirect product of $Z/3Z$ with normal subgroup V. If the preimage of V in J' is torsion free, then by the calculations of Theorem 4 we may assume that V acts through the subgroup generated by $t = diag(1,-1,-1)$ and $u = diag(-1,1,-1)$ in $SL(3,Z)$. (This part of the argument did not involve consideration of meridianal actions). If w is an integral 3×3 matrix of order 3 such that $wt = uw$ and $wu = tuw$ then we may solve these linear equations for the entries of w , and we find that (up to multiplication by 1, t, u or tu) w is a permutation matrix. The group J' must then have a presentation of the form

$$<Z^3,t,u,w \mid tx = xt,\ tyt^{-1} = y^{-12},\ tzt^{-1} = z^{-1},\ uxu^{-1} = x^{-1},\ uy = yu,$$

$$uzu^{-1} = z^{-1},\ wxw^{-1} = y,\ wyw^{-1} = z,\ wzw^{-1} = x,\ u = wtw^{-1},$$

$$t^2 = x^a y^b z^c,\ w^3 = x^d y^e z^f,\ (tw)^3 = x^g y^h z^i>$$

for some $a, \cdots i$ in Z. But the abelianization of such a group is $Z/6Z$, and so cannot admit a meridianal automorphism. Therefore no such J' can be torsion free. □

Similar calculations show that there is no subgroup of $SL(3,Z)$ isomorphic to A_4 whose Sylow 2-subgroup is as in the third case treated in Theorem 4, while there are several in each of the other two cases.

Abelian normal subgroups of rank greater than 2

In this section we shall determine all the 2-knot groups with large abelian normal subgroups.

Theorem 6 *Let π be a 2-knot group with an abelian normal subgroup of rank greater than 2. Then π' is isomorphic to Z^3 or G_6.*

Proof Let A be a maximal abelian normal subgroup of π. Then zA is also a normal subgroup and $J = \pi/zA$ is a PD_4^+-group over Q, by the results of Chapter 3. Hence the rank of A/zA (i.e. the rank of A) is at most 4. Moreover if A has rank 4 then $[\pi:A] = [J:A/zA]$ is finite, and so A is finitely generated. Hence π has a torsion free abelian normal subgroup of the same rank and so is in fact a virtually abelian PD_4^+-group. If A has rank 3 then A/zA must have infinite index in J, so $c.d.A/zA = 3 =$ rank A/zA and therefore A/zA is finitely generated [B: Theorem 7.14]. Now π must have an element, g say, whose image in $\pi/A = J/(A/zA)$ has infinite order, by [B: Theorem 8.4]. (Note that the accessibility criterion used in this theorem has since been extended to arbitrary coeffficient rings [Du 1985]). The subgroup of J generated by the images of A and g has Hirsch length 4, and so must have finite index there. Therefore the subgroup B of π generated by A and g has finite index, and so is finitely presentable. The group A is thus finitely generated as a module over $Z[B/A] \cong \Lambda$ and so its torsion subgroup has finite exponent. Therefore π again has a torsion free abelian normal subgroup of the same rank and so is again a PD_4^+-group. The last assertion now follows from Theorem 5. □

It remains for us to determine the possible meridianal automorphisms of the commutator subgroup.

Theorem 7 *Let π be a 2-knot group with $\pi' \cong Z^3$. Then the meridianal automorphism of π' is given by a matrix C in $SL(3,Z)$ such that $|det(C-I)| = 1$. The characteristic polynomial of C is irreducible, and determines π up to a finite ambiguity. Moreover π has no abelian normal subgroup of rank greater than 3, and $\zeta\pi$ is trivial.*

Proof Let K be a 2-knot with group π and let t be a meridian for π. Let C be the matrix in $GL(3,Z)$ of the action of t by conjugation on $\pi' \cong Z^3$. Then $t-1$ acts invertibly, so $det(C-I) = \pm 1$. Moreover $M(K)$ is orientable and aspherical, so $det(C) = +1$. Since the characteristic polynomial of C has integral coefficients, leading coefficient and constant term 1 and does not vanish at ± 1, it must be irreducible. By a theorem of Latimer and MacDuffee the conjugacy classes of matrices in $GL(n,Z)$ with given irreducible characteristic polynomial correspond to the ideal classes of the number field generated by a root of the polynomial [New: page 52]. Therefore the group π is determined up to a finite ambiguity by its Alexander polynomial. Moreover the characteristic polynomial of C cannot be cyclotomic, and so no power of t can be central. The last assertion follows easily. □

Every such matrix is the meridianal automorphism of some such 2-knot group, by Theorem 9 of Chapter 2. In the appendix to [AR 1984] there are several examples in which the conjugacy class is uniquely determined by the characteristic polynomial.

Theorem 8 *Let π be a 2-knot group with $\pi' \cong G_6$. Then $\pi \cong G(+)$ or $G(-)$, where $G(e)$ is the group with presentation*

$$\langle x,y,t \mid xy^2x^{-1}y^2 = 1,\ txt^{-1} = (xy)^{-e},\ tyt^{-1} = x^e \rangle$$

for $e = \pm 1$. In each case $\zeta\pi$ is infinite cyclic and $\pi' \cap \zeta\pi = 1$.

Proof We must find the conjugacy classes in $Out(G_6)$ which contain meridianal automorphisms. The group G_6 has a presentation

$$\langle x,y,z \mid xy^2x^{-1}y^2 = yx^2y^{-1}x^2 = 1,\ z = xy \rangle.$$

The subgroup A_6 generated by $\{x^2, y^2, z^2\}$ is a maximal abelian normal subgroup, isomorphic to Z^3, and $G_6/A_6 \cong V$. Define automorphisms i and j of G_6 by $i(x) = y$, $i(y) = x$ (hence $i(z) = x^{-2}y^2z^{-1}$ and $i^2 = id$) and $j(x) = xy$, $j(y) = x$ (hence $j(z) = xyx = z^2y^{-1}$ and $j^6 = id$). Then the images of i and j generate $Aut(G_6/A_6) \cong GL(2,F_2)$. Let E be the subgroup of $Aut(G_6)$ generated by the automorphisms α, β, γ, δ, ε and ϕ sending the generator x to x^{-1}, x, x, x, y^2x and z^2x respectively, and the generator y to y, y^{-1}, z^2y, x^2y, y and z^2y respectively. (Hence the generator z is sent to $x^{-2}z$, y^2z, z^{-1}, x^2z, y^2z and z respectively. Note that these automorphisms act on A_6 via diagonal matrices, with respect to the basis $\{x^2, y^2, z^2\}$).

Then $E = ker(:Aut(G_6) \rightarrow Aut(G_6/A_6))$. For an automorphism inducing the identity on G_6/A_6 must send x to $x^{2p}y^{2q}z^{2r}x$, y to $x^{2s}y^{2t}z^{2u}y$ and hence z to $x^{2(p+s)}y^{2(q-t)}z^{2(r-u)+1}$. The squares of these elements are x^{4p+2}, y^{4t+2} and $z^{4(r-u)+2}$, which generate A_6 if and only if $p = -1$ or 0, $t = -1$ or 0 and $r = u-1$ or u. Composing such an automorphism appropriately with α, β and γ we may achieve $p = t = 0$ and $r = u$. Then by composing with powers of δ, ε and ϕ we may obtain the identity automorphism. These automorphisms satisfy $\alpha^2 = \beta^2 = \gamma^2 = 1$ and each pair commutes except for $\alpha\delta = \delta^{-1}\alpha$, $\beta\varepsilon = \varepsilon^{-1}\beta$ and $\gamma\phi = \phi^{-1}\gamma$. The inner automorphisms are contained in E, and are generated by $\beta\gamma\delta$ (conjugation by x) and $\alpha\gamma\varepsilon\phi$ (conjugation by y). Therefore $\overline{E} = E/Inn(G_6)$ is a group of exponent 2 generated by the images of α, β, γ and ε.

Since the images of these elements in $Aut(G_6/G_6') \cong GL(2,Z/4Z)$ are $\begin{pmatrix}-1 & 0\\ 0 & 1\end{pmatrix}, \begin{pmatrix}1 & 0\\ 0 & -1\end{pmatrix}, \begin{pmatrix}1 & 2\\ 0 & -1\end{pmatrix}$ and $\begin{pmatrix}1 & 0\\ 2 & 1\end{pmatrix}$ respectively (with respect to the basis given by the images of x and y) and these matrices generate a group of order 16, we must have $\overline{E} = (Z/2Z)^4$. Since in $Aut(G_6)$ we have $j^3 = \alpha\beta\gamma\varepsilon$, $jiji = \delta$, $i\alpha = \beta i$, $\phi i\gamma = \gamma i$, $j\alpha = \gamma j$, $j\beta = \delta\alpha j$, $j\gamma = \beta\varepsilon j$ and $j\varepsilon = \delta j$, we find that $Out(G_6)$ is a group of order 96, with a presentation

$$<i,j,\alpha,\beta,\gamma,\varepsilon \mid \alpha^2 = \beta^2 = \gamma^2 = \varepsilon^2 = i^2 = j^6 = 1,\ \alpha,\beta,\gamma,\varepsilon \textit{ commute},\ i\alpha = \beta i,$$
$$i\gamma = \alpha\varepsilon i,\ j\alpha = \gamma j,\ j\beta = \alpha\beta\gamma j,\ j\gamma = \beta\varepsilon j,\ j\varepsilon = \beta\gamma j,\ j^3 = \alpha\beta\gamma\varepsilon,\ jiji = \beta\gamma>.$$

If ρ is a meridianal automorphism of G_6, then it must induce a meridianal automorphism of G_6/A_6, and so we must have $\rho \equiv j$ or j^{-1} modulo E. Conversely any such automorphism is meridianal, for it implies that G_6 modulo $<<g^{-1}\rho(g) \mid g \textit{ in } G_6>>_{G_6}$ is a perfect group, and therefore is trivial as G_6 is solvable. There are 32 elements in the cosets $j\overline{E}\cup j^{-1}\overline{E}$ of $Out(G_6)$. The centralizer of j in $Out(G_6)$ is generated by $\alpha\beta$ and j, and has order 12. The distinct cosets of this centralizer in $Out(G_6)$ are represented by $\{1, \alpha, \gamma, \alpha\gamma, i, i\alpha, i\gamma, i\alpha\gamma\}$. Conjugating j and j^{-1} by these elements we get 16 distinct elements of $j\overline{E}\cup j^{-1}\overline{E}$, which all give rise to the group with presentation

$$<x,y,t \mid xy^2x^{-1}y^2 = 1,\ txt^{-1} = xy,\ tyt^{-1} = x>.$$

However this group cannot be a PD_4^+-group, as already the subgroup generated by $\{x^2, y^2, z^2, t\}$ is nonorientable. The elements $j\alpha$ and $j\beta$ also have centralizers of order 12 and their conjugates exhaust the remaining 16 elements of $j\overline{E}\cup j^{-1}\overline{E}$. Each of $j\alpha$ and $j\beta$ is conjugate to its inverse (via i), and so the groups $G(+)$ and $G(-)$ that they give rise to are distinct. Moreover these automorphisms are orientation preserving on A_6 and hence on G_6 (in fact $j\alpha = (j\alpha)^4$) and so $G(+)$ and $G(-)$ are PD_4^+-groups.

In each case A_6 is an abelian normal subgroup of rank 3, while the subgroup generated by $A_6\cup\{t^6\}$ is an abelian normal subgroup of rank 4. Since the characteristic polynomial of t acting on A_6 is X^3-1, the only candidates for normal subgroups of rank less than 3 contained in A_6 are (essentially) $(t-1)A_6$, generated by $\{x^2y^2, x^2z^{-2}\}$ and $(t^2+t+1)A_6$, generated by $\{x^2y^{-2}z^2\}$. It is easy to see that neither of these groups is even normal in $G(e)'$. Therefore any abelian normal subgroup B of $G(e)$ such that $A_6\cap B$ has rank less than 3 must map injectively to the abelianization, and so have rank 1. Such a group must be central. In fact $\zeta G(+)$ and $\zeta G(-)$ are generated by t^3 and $t^6x^{-2}y^2z^{-2}$ respectively. This completes the proof of the theorem. □

The group $G(+)$ is the group of the 3-twist spin of the figure eight knot. On the other hand, it can be shown that no power of a weight element is central in $G(-)$ and so it is not the group of any twist spin, although it is the group of a fibred 2-knot, by Theorem 9 of Chapter 2.

Nilpotent HP radical

We may try to analyse π'/T in a similar fashion when it has Hirsch–Plotkin radical isomorphic to Γ_q for some q.

Theorem 9 *Let J be a PD_4^+-group over Q with $J/J' \cong Z$ and which has no nontrivial torsion normal subgroup. If the Hirsch–Plotkin radical H of J' is isomorphic to Γ_q for some $q \geqslant 1$ then either $J' = H$ or the quotient J'/H is isomorphic to $Z/3Z$ or V. In particular, J is solvable, of derived length at most 4.*

Proof Let j be an element of J which is not in J'. Since H is normal in J, the subgroup generated by H and j is a poly-Z group of Hirsch length 4, and so has homological dimension 4. Therefore it has finite index in J (by [B: Theorem 9.22]) and so J'/H is finite.

Let $\alpha: J \rightarrow Aut(H/\zeta H)$ be the homomorphism determined by the conjugation action of J on H. Then H is contained in $K = J' \cap ker(\alpha)$ and has finite index there. Since $\zeta H \cong Z$, it is central in K and as $H/\zeta H$ is central in $K/\zeta H$ we may apply Baer's extension of Schur's theorem [R: 14.5.1] to conclude that $[K,K']$ is finite. Since it is normal in J, it must be trivial. Therefore K is a nilpotent normal subgroup of J' and so $K = H$ (by maximality of H). Since $\alpha(J')$ is a finite subgroup of $Aut(H/\zeta H) \cong GL(2,Z)$, it is cyclic or dihedral, of order dividing 24. Since it is the commutator subgroup of J/K it must admit a meridianal automorphism, and so must be trivial, $Z/3Z$ or V. □

We shall assume henceforth that π is torsion free. Then π' is the fundamental group of a closed orientable Seifert fibred 3-manifold with a *Nil*-structure, by [Sc 1983] and [Z: Section 63]. (Recall that *Nil* is the nilpotent Lie group of 3×3 upper triangular real matrices). These are either circle bundles over the torus or have base S^2 and 3 or 4 exceptional fibres, of type (α_i,β_i), with $\Sigma\alpha_i^{-1} = 1$. Each circle bundle over the torus is the coset space of *Nil* by a discrete uniform subgroup isomorphic to Γ_q, for some $q \geqslant 1$. We shall treat these cases first.

If ϕ is an automorphism of Γ_q, sending x to $x^a y^b z^m$ and y to

$x^c y^d z^n$ for some $a, \cdots n$ in Z then it must send z to z^{ad-bc}. The integral matrix $A = \begin{pmatrix} a & c \\ b & d \end{pmatrix}$ represents the induced automorphism of $\Gamma_q/\zeta\Gamma_q = Z^2$, and every pair (A,μ) with A in $GL(2,Z)$ and $\mu = (m,n)$ in Z^2 determines an automorphism of Γ_q. The multiplication rule for such ordered pairs is somewhat messy: $(A,\mu)(B,\nu) = (AB,\mu B+(detA)\nu+q\,\omega(A,B))$ where $\omega(A,B)$ is biquadratic in the entries of A and B. The map p from $Aut(\Gamma_q)$ to $GL(2,Z)$ which sends (A,μ) to A is a homomorphism, and $ker(p) \cong Z^2$. On factoring out the inner automorphisms, represented by $q.ker(p)$, we find that $Out(\Gamma_q)$ is a semidirect product of $GL(2,Z)$ with the normal subgroup $(Z/qZ)^2$, with multiplication given by $[A,\mu][B,\nu] = [AB,\mu B+(detA)\nu]$, where $[A,\mu]$ is the image of (A,μ) in $Out(\Gamma_q)$. In particular, $Out(\Gamma_1) = GL(2,Z)$.

Theorem 10 *Let π be a 2-knot group with $\pi' \cong \Gamma_q$. Then q is odd and the image of the meridianal automorphism in $Out(\Gamma_q)$ is conjugate to $[A,0]$, where A is one of $\begin{pmatrix} 1 & 1 \\ 1 & 2 \end{pmatrix}$, $\begin{pmatrix} 1 & -1 \\ 1 & 0 \end{pmatrix}$, $\begin{pmatrix} 1 & 1 \\ 1 & 0 \end{pmatrix}$ or $\begin{pmatrix} 0 & 1 \\ 1 & -1 \end{pmatrix}$. If $q > 1$ only the latter two matrices determine meridianal automorphisms. (Moreover they represent mutually inverse automorphisms).*

Proof If (A,μ) is meridianal then so are the induced automorphisms of $\Gamma_q/\zeta\Gamma_q = Z^2$ and of $z(\Gamma_q/\Gamma_q') = Z/qZ$. Therefore $|det(A-I)| = 1$ and $detA-1$ is a unit modulo (q), so q must be odd, and $detA = -1$ if $q > 1$. The characteristic polynomial of such a 2×2 matrix must be one of X^2-3X+1, X^2-X+1, X^2-X-1 or X^2+X-1. The corresponding fields $Q(\sqrt{5})$ and $Q(\sqrt{-3})$ each have class number 1 and so the conjugacy class of such a matrix is determined by its characteristic polynomial [New: page 52]. Thus A must be conjugate to one of the above matrices. The inverse of $[A,\mu]$ is $[A^{-1},-(detA)\mu A^{-1}]$, so $[A,\mu][A,\nu][A,\mu]^{-1} = [A,\mu+(detA)(\nu-\mu)A^{-1}] = [A,\mu(A-(detA)I)A^{-1}+(detA)\nu A^{-1}]$. Since $A-(detA)I$ is invertible, $[A,\nu]$ is conjugate to $[A,0]$. (The remark in parentheses now follows as the latter two of the given matrices are mutually inverse). □

Every automorphism (A,μ) of Γ_q is orientation preserving. This follows, with some effort, from the criterion of [B: page 177]. Alternatively, any self homotopy equivalence of a nontrivial S^1-bundle over an orientable surface of positive genus is orientation preserving [Fr 1975]. Each of the groups allowed by Theorem 10 is the group of some fibred 2-knot, by Theorem 9 of Chapter 2. We shall see in Chapter 8 that, except when $q = 1$ and $detA$ is positive, the knot realising such a group is unique up to changes of orientation. The group of the 6-twist spin of the trefoil has commutator subgroup Γ_1. In all the other cases the meridianal automorphism has infinite order and the group is not the group of any twist spin.

Among the Nil-manifolds which are Seifert fibred with base S^2 only those with 3 exceptional fibres of type $(3,\beta_i)$, with $\beta_i=\pm 1$, are of interest to us, as in all other cases the quotient of the fundamental group by its Hirsch-Plotkin radical is cyclic of even order, and so the fundamental group does not admit a meridianal automorphism. (In particular, there is no 2-knot group π with π' an extension of V by some Γ_q). Such 3-manifolds are 2-fold cyclic branched covers of S^3, branched over a Montesinos link $K(0\,|\,b;(3,\beta_1),(3,\beta_2),(3,\beta_3))$ [Mo 1973]. Up to change of orientation, we may assume that $\beta_1 = \beta_2 = 1$. If b is odd then the first homology does not admit a meridianal automorphism. Therefore b must be even, and so the link is a knot. These knots are all invertible, but none are amphicheiral [BZ: Chapter 12E]. (Among them are the knots 9_{35}, 9_{37}, 9_{46}, 9_{48}, 10_{74} and 10_{75}).

Let $\pi(b,\varepsilon)$ be the group of the 2-twist spin of $K(0\,|\,b;(3,1),(3,1),(3,\varepsilon))$.

Theorem 11 *Let π be a 2-knot group such that π' is a (torsion free) extension of $Z/3Z$ by Γ_q, for some $q \geqslant 1$. Then q is odd, and π is isomorphic to $\pi(q+2+\varepsilon,\varepsilon)$, for some $\varepsilon = \pm 1$.*

Proof By the remarks above, we may assume that π' is the fundamental group of the 2-fold cyclic branched cover of $K(0\,|\,b;(3,1),(3,1),(3,\varepsilon))$ for some even b and $\varepsilon = \pm 1$. Therefore it has a presentation of the form

$$<r,s,t,h \mid r^3 = s^3 = t^{3\varepsilon} = h,\ rst = h^b>.$$

A Reidemeister–Schreier calculation shows that the subgroup of index 3 normally generated by the images of rs^{-1}, rt^{-1} and h is isomorphic to $\Gamma_{b-2-\varepsilon}$. Hence $b = q+2+\varepsilon$, and q is odd. The centre of π' is generated by the image of h, and $G = \pi'/<h>$ has a presentation $<r,s \mid r^3 = s^3 = (rs)^3 = 1>$. If we let $u = r^{-1}s$ and $v = sr^{-1}$ we see that G also has a presentation

$$<r,u,v \mid r^3 = 1,\ uv = vu,\ rur^{-1} = v,\ rvr^{-1} = u^{-1}v^{-1}>$$

and thus is an extension of $Z/3Z$ by $<<u,v>> \cong Z^2$. Since π'/π'' is finite, $Hom(\pi',\zeta\pi') = 1$ and so the natural map from $Out(\pi')$ to $Out(G)$ is injective. We shall show that it is an isomorphism, and that all meridianal automorphisms of π' are conjugate in $Out(\pi')$.

Let $C = \begin{pmatrix}1 & -1\\ 1 & 0\end{pmatrix}$, $J = \begin{pmatrix}0 & 1\\ 1 & 0\end{pmatrix}$ and $R = \begin{pmatrix}0 & -1\\ 1 & -1\end{pmatrix}$. Then $C^2 = R$, and the normalizer of $<R>$ in $GL(2,Z)$ is the subgroup $<C,J>$ generated by C and J, which is dihedral of order 12. The automorphism of $<<u,v>>$ determined by conjugation by r has matrix R. It is not hard to see that each automorphism of G corresponds to a pair (A,μ) in $GL(2,Z)\times Z^2$ such that $ARA^{-1} = R^{detA}$. (Thus A must be in the dihedral subgroup $<C,J>$). The automorphism $(\begin{pmatrix}a & c\\ b & d\end{pmatrix},(m,n))$ sends t to $t^{ad-bc}u^m v^n$, u to $u^a v^b$ and v to $u^c v^d$, and the multplication in $Aut(G)$ is given by $(A,\mu)(B,\nu) = (AB,A\nu+(detB)R^{\delta detA}\mu)$, where $\delta = ½(detB-1)$. In fact, $Aut(G)$ is a semi-direct product of the dihedral group generated by $(C,(0,0))$ and $(J,(0,0))$ with the normal subgroup consisting of all pairs (I,μ), which is isomorphic to Z^2. The subgroup of inner automorphisms is generated by $c_r = (R,(0,0))$, $c_u = (I,(-2,-1))$ and $c_v = (I,(1,-1))$. It then follows that $Out(G)$ has a presentation

$$<c,j,k \mid c^2 = j^2 = k^3 = 1,\ cj = jc,\ ckc^{-1} = k^{-1},\ jkj^{-1} = k^{-1}>$$

where c, j and k represent the classes of $(C,(0,0))$, $(J,(0,0))$ and $(I,(1,0))$ repectively, and that the natural map from $Aut(\pi')$ to $Out(G)$ is onto. On considering the effect of an automorphism of π' on its characteristic quotients $\pi'/H = G/<<u,v>> \cong Z/3Z$ and $G/G' = (Z/3Z)^2$, we see that the only outer automorphism classes which contain meridianal automorphisms are j, jk and jk^2. Since these are conjugate in $Out(G)$ and $\pi' \cong \pi(b,1,1,\varepsilon)'$, the theorem now follows from Lemma 1 of Chapter 2. □

Virtually solvable groups

By Theorem 6 of Chapter 3 and Theorems 1 and 2 above we may assume that the quotient of a virtually solvable 2-knot group by its maximal locally-finite normal subgroup is either Z, Φ or is a virtually poly-Z group of Hirsch length 4. Moreover in the first case the commutator subgroup is finite, and all such groups are listed in Theorem 3 of Chapter 4. Our results for the other cases are not yet complete.

Lemma 3 *Let G be a finitely presentable group with a torsion normal subgroup T such that G/T is either virtually poly-Z or is a finite extension of Φ. Then T/T' is finitely presentable as a $Z[G/T]$-module, and in particular has finite exponent as an abelian group.*

Proof Let C be a finite 2-complex with fundamental group G. Then $T/T' = H_1(C;Z[G/T])$. In the first case $Z[G/T]$ is noetherian [R: 15.3.3], while in the second case it is coherent [BS 1979]. In each case the homology of a finitely generated free $Z[G/T]$-chain complex is finitely presentable. If T/T' is generated as a $Z[G/T]$-module by elements t_i of order e_i then Πe_i is a finite exponent for T/T'. □

This lemma suggests that there may be a homological proof that solvable 2-knot groups are virtually torsion free. For if π is solvable then so is its maximal locally-finite normal subgroup T, and so T is trivial if and only if $T/T' = 0$. As T/T' has finite exponent, T is trivial if and only if $H_1(T;F_p) = 0$ for all primes p. Note also that $F_p[\Phi]$ is a coherent Ore domain of global dimension 2, while if J is a poly-Z group of Hirsch length 4 then $F_p[J]$ is a noetherian Ore domain of global dimension 4 (cf. [P: 4.4, 13.3]).

Theorem 12 *Let π be a 2-knot group which has a locally-finite normal subgroup T such that $\pi/T \cong \Phi$. Then $H^2(\pi;Q[\pi]) \neq 0$.*

Proof Since T is locally-finite, $h.d._Q T = 0$ and so $h.d._Q \pi = 2$. Therefore the $Q[\pi]$-submodule B_1 of 1-boundaries in the cellular chain complex C_* of $\tilde{M}$ with coefficients Q is projective, and so the submodule Z_2 of 2-cycles is a direct summand of C_2. Hence $\Pi = H_2(\tilde{M};Q)$ is finitely presentable as a $Q[\pi]$-module. By the Universal Coefficient spectral sequence and Poincaré duality in M there is a homomorphism d from Π to $\overline{Hom_{Q[\pi]}(\Pi,Q[\pi])}$ with kernel $\overline{H^2(\pi;Q[\pi])}$.

Since $h.d._Q T = 0$ tensoring with $Q[\pi/T]$ is exact, and so the Cartan-Leray spectral sequence from $\tilde{M}$ to M_T (the covering space with group T) collapses. Therefore $\Pi_T = H_2(M_T;Q)$ is isomorphic to $\Pi\otimes Q[\pi/T]$. Moreover the natural map from $\overline{Hom_{Q[\pi]}(\Pi,Q[\pi])\otimes Q[\pi/T]}$ to $\overline{Hom_{Q[\pi/T]}(\Pi_T,Q[\pi/T])}$ is an isomorphism, since Π is finitely presentable. The homomorphism from Π_T to the latter module induced by d may be identified with the one given by the Universal Coefficient spectral sequence and Poincaré duality in M_T, which has kernel $\overline{H^2(\pi/T;Q[\pi/T])}$. Since $\pi/T \cong \Phi$ this kernel is nonzero, and so the theorem follows from the exactness of $-\otimes Q[\pi/T]$. □

If such a group is solvable it is an ascending HNN extension with finitely generated base [BS 1978]. The base is then an extension of Z by a locally-finite normal subgroup, which cannot be finite, for otherwise T would be finite. Therefore the base has one end. The above theorem together with Theorem 0.1 of [BG 1985] then imply that the base cannot be almost finitely presentable. It seems unlikely that there are any such 2-knot groups.

Corollary *If π also has an abelian normal subgroup A with an element of infinite order then $\pi \cong \Phi$.*

Proof Clearly A has rank 1, and $A\cap T = zA$. If T/zA were infinite then π/A would have one end, and so $H^2(\pi;Q[\pi]) = 0$ [Mi 1986], contradicting the theorem. Therefore π/zA is a finite extension of Φ, and so zA has finite exponent, e say, by Lemma 3. Since eA is then a torsion free rank 1 abelian normal subgroup of π the same argument shows that T must be finite, and hence trivial by Theorem 6 of Chapter 4. □

The argument for our next theorem depends on an unpublished result of Kropholler on solvable PD-groups over Q.

Theorem 13 *Let π be a virtually solvable 2-knot group with maximal locally-finite normal subgroup T. Suppose that π/T is a virtually poly-Z group of Hirsch length 4. If π also has an abelian normal subgroup A of rank > 0, then it is a solvable PD_4^+-group, and π' is virtually nilpotent.*

Proof Clearly $A \cap T = zA$, so A/zA is isomorphic to a torsion free subgroup of π/T, and is thus free abelian. Theorems 3, 4 and 5 of Chapter 3 then imply that π/zA is a PD_4^+-group over Q. Since it is virtually solvable it must be virtually poly-Z [Kr 1987], and so zA has finite exponent, e say, by Lemma 3. Since eA is a free abelian normal subgroup of π of rank > 0, the theorem follows from our earlier work. □

When π' is virtually nilpotent no assumption on abelian normal subgroups is needed, and the classification can be made explicit.

Theorem 14 *Let π be a 2-knot group with virtually nilpotent commutator subgroup. Then either $\pi \cong \Phi$, $G(+)$, $G(-)$ or $\pi(b,\varepsilon)$, for some even b and $\varepsilon = \pm 1$, or $\pi' \cong Z^3$ or Γ_q, for some odd q, or π' is finite. In the latter cases the meridional automorphisms are as in Theorems 7 and 10 of this Chapter or as in Theorem 3 of Chapter 4.*

Proof By the remark on page 265 of [BS 1978], π is constructible and so is virtually torsion free [BB 1976]. The theorem now follows from our earlier work. □

If every virtually solvable 2-knot group with infinite commutator subgroup is torsion free then Theorem 14 gives *all* the 2-knot groups with ascending series whose factors are either locally-finite or locally-nilpotent.

Chapter 7 THE HOMOTOPY TYPE OF $M(K)$

In the remaining two chapters we shall attempt to characterize certain 2-knots in terms of algebraic invariants. As we showed in Chapter 1 we may recover a 2-knot K (up to orientations and Gluck reconstruction) from $M(K)$ together with a weight class in πK. Thus we seek to determine M up to homeomorphism. This problem breaks up naturally into two parts. First we find the homotopy type and then we attempt to apply surgery to classify the manifolds within that homotopy type.

In this chapter we shall consider the homotopy classification. Our results are consistent with the conjecture that when π' is finitely generated then M' is a PD_3^+-complex, and may thus be regarded as homotopy theoretic approximations to a 4-dimensional fibration theorem. We prove this in two cases: when $\zeta\pi$ is not contained in π' and when π' is free. (These include the groups of twist spins of 1-knots, and of 0-spins of fibred 1-knots, respectively). We suppose first that π' is finite, and show that the homotopy type is determined by the group and the first nontrivial k-invariant. After an algebraic interlude we give our main results, treating the two cases together as far as possible. (See [Hi 198?] for an extension to the nonorientable cases).

We consider also 2-knots whose groups have finite geometric dimension 2 and one end. (Among these are 1-knot groups, knot groups with free commutator subgroup and Φ). If K and K_1 are two such knots with isomorphic groups then there is a 3-connected map from $M(K_1)-int\,D^4$ to $M(K)$. If the map extends across the deleted 4-cell then the extension is a homotopy equivalence. The group determines $\pi_2(M)$ and equivariant Poincaré duality up to sign, but we do not know how duality constrains the attaching map for the 4-cell.

If a 2-knot K has a Seifert hypersurface V such that the "pushoff" maps on either side into its complement in X each induce monomorphisms on the fundamental groups then M' is homotopy equivalent to the closed 3-manifold $\hat{V}$ obtained by capping off ∂V with a disc. If moreover every self homotopy equivalence of $\hat{V}$ is homotopic to a homeomorphism then M is homotopy equivalent to $M(K_1)$, where K_1 is a fibred 2-knot with closed fibre $\hat{V}$.

The spherical case

We saw in Theorem 1 of Chapter 4 that if π' is finite then the universal cover $\widetilde{M}$ is homotopy equivalent to S^3. (The quotient $M' = \widetilde{M}/\pi'$ is then a PD_3^+-complex [Wa 1967]).

Theorem 1 *Let K be a 2-knot with group π such that π' is finite. Then the homotopy type of M is determined by the group π and the class of the first nontrivial k-invariant, which is a generator k_4 of $H^4(\pi;Z) \cong Z/|\pi'|Z$, modulo the action of $Aut(\pi)$ and $Aut(Z)$.*

Proof Since $\pi_2(M) = 0$ and $\pi_3(M) \cong Z$, by Theorem 1 of Chapter 4, the first nontrivial k-invariant is k_4 in $H^4(\pi;Z)$. By examining the effect of the meridianal automorphism on the Sylow subgroups of π', we see that π acts trivially on $H^4(\pi';Z)$. By the LHS spectral sequence for π as an extension of Z by π', we then have $H^4(\pi;Z) = H^4(\pi';Z)$, and by the Universal Coefficient theorem this is isomorphic to $H_3(\pi';Z) \cong Z/|\pi'|Z$. Moreover k_4 clearly also represents the first k-invariant of the PD_3-space M', and so must be a generator [Wa 1967].

Now let N be the third stage of the Postnikov tower for M and let $j:M \to N$ be the natural map. Then j is 4-connected and we may construct N by adjoining cells of dimension greater than 4 to M. If K_1 is another 2-knot and $f:\pi K_1 \to \pi K$ is an isomorphism which identifies the k-invariants up to sign, then there is a 4-connected map $j_1:M(K_1) \to N$ inducing f, which is homotopic to a map with image in the 4-skeleton of N, and so there is a map $h:M(K_1) \to M(K)$ such that j_1 is homotopic to jh. The map h induces isomorphisms on π_i for $i \leqslant 3$, since j and j_1 are 4-connected, and so the lift $\tilde{h}:\widetilde{M}_1 \sim S^3 \to \widetilde{M} \sim S^3$ is a homotopy equivalence, by the theorems of Hurewicz and Whitehead. It follows that h is itself a homotopy equivalence. □

Examples in which π' is odd cyclic but π alone does not determine the homotopy type of M can be found among the 2-twist spins of 2-bridge knots (cf. Theorem 6.2 of [Pl 1983]).

We shall need to know something about the self homotopy equivalences of M when we make use of the exact sequence of surgery, in Chapter 8.

Theorem 2 *Let K be a 2-knot with group π such that π' is finite. Then the group of self homotopy equivalences of $M(K)$ is finite.*

Proof If f is a self homotopy equivalence of $M(K)$ then we may assume that it fixes a base point. Since $Aut(\pi')$ is finite, we may then assume that f is orientation preserving and induces the identity on π. Let M_ζ be the (finite) covering space of M associated with the subgroup $\zeta\pi$. Since f lifts to self homotopy equivalences f_ζ and $\tilde{f}$ of the covering spaces M_ζ and $\tilde{M}$, and since f_ζ is orientation preserving, it follows that $\tilde{f}$ induces the identity on $H_3(\tilde{M};Z)$. Therefore it is homotopic to $id_{\tilde{M}}$, since $\tilde{M} \sim S^3$, and so induces the identity on all homotopy groups. We may assume that M has a finite cell structure with a single 4-cell. There are no obstructions to constructing a homotopy from f to id_M on the 3-skeleton $M_o = M - int\, D^4$, and since $\pi_4(M) = \pi_4(S^3) = Z/2Z$ there are just two possibilities for f. □

Cup products

In this section we shall adapt and extend the work of Barge [Ba 1980,1980'] in setting up duality maps in the equivariant cohomology of covering spaces. Let G be a group with a normal subgroup H such that the quotient G/H is infinite cyclic, and fix an element t in G whose image generates G/H. Let $\alpha:H \to H$ be the automorphism determined by $\alpha(h) = tht^{-1}$ for all h in H. Let $\Gamma = Z[G]$, $\Theta = Z[H]$ and $\Lambda = Z[t,t^{-1}]$. The automorphism α of H extends to a ring automorphism of Θ, and the ring Γ may then be viewed as a twisted Laurent extension, $\Gamma = \Theta_\alpha[t,t^{-1}]$. The ring Λ is the quotient of Γ by the two-sided ideal generated by $\{h-1 \mid h \text{ in } H\}$, while as a left module over itself Θ is isomorphic to $\Gamma/\Gamma(t-1)$ and so may be viewed as a left Γ-module. (Note that α is not a module automorphism unless t is central). Unless otherwise stated, all modules over these rings shall be left modules.

If M is a (left) Γ-module let $\ddot{M}$ denote the underlying Θ-module, and let $\hat{M} = Hom_{\Theta}(\ddot{M},\Theta)$. Then $\hat{M}$ is a *right* Θ-module via $(f\theta)(m) = f(m)\theta$ for all θ in Θ, f in $\hat{M}$ and m in M. If $M = \Gamma$ then $\hat{\Gamma}$ is also a left Γ-module via $(\theta t^r f)(\phi t^s) = \phi\alpha^{-s}(\theta)f(t^{s-r})$ for all f in $\hat{\Gamma}$, ϕ, θ in Θ and r, s in Z. As the left and right actions commute, $\hat{\Gamma}$ is a (Γ,Θ)-bimodule. We may describe this bimodule more explicitly. Let $\Theta[[t,t^{-1}]]$ be the set of doubly infinite power series $\Sigma t^n\theta_n$ with θ_n in Θ for all n in Z, with the obvious right Θ-module structure and with the left Γ-module structure given by $\theta t^r(\Sigma t^n\theta_n) = \Sigma t^{n+r}\alpha^{-n-r}(\theta)\theta_n$ for all θ, θ_n in Θ and r in Z. (Note that even if $H = 1$, so $\Theta = Z$, this module is not a ring in any natural way). Then the map $j:\hat{\Gamma} \to \Theta[[t,t^{-1}]]$ given by $j(f) = \Sigma t^n f(t^n)$ for all f in $\hat{\Gamma}$ is a (Γ,Θ)-bimodule isomorphism.

Given f in $\hat{M}$ we may define a map $T_M f:M \to \hat{\Gamma}$ by $(T_M f)(m)(t^n) = f(t^{-n}m)$ for all m in M and n in Z. It is routine to check that $T_M f$ is Γ-linear, and that $T_M:\hat{M} \to Hom_{\Gamma}(M,\hat{\Gamma})$ is an isomorphism of abelian groups. (It is clearly a monomorphism, and if $g:M \to \hat{\Gamma}$ is Γ-linear then $g = T_M f$ where $f(m) = g(m)(1)$ for all m in M. In fact if we give $Hom_{\Gamma}(M,\hat{\Gamma})$ the natural right Θ-module structure by $(\mu\theta)(m) = \mu(m)\theta$ for all θ in Θ, Γ-linear maps $\mu:M \to \hat{\Gamma}$ and m in M then T_M is an isomorphism of right Θ-modules). Thus we have a natural equivalence $T:Hom_{\Theta}(\ddot{-},\Theta) \to Hom_{\Gamma}(-,\hat{\Gamma})$ of functors from $((\Gamma\text{-}Mod))$ to $((\Phi\text{-}Mod))$. If C_* is a Γ-chain complex then T induces natural isomorphisms from $H^*(\ddot{C}_*;\Theta) = H^*(Hom_{\Theta}(\ddot{C}_*,\Theta))$ to $H^*(C_*;\hat{\Gamma}) = H^*(Hom_{\Gamma}(C_*,\hat{\Gamma}))$. In particular since the forgetful functor $\ddot{-}$ is exact and takes projectives to projectives there are isomorphisms from $Ext^*_{\Theta}(\ddot{M},\Theta)$ to $Ext^*_{\Gamma}(M,\hat{\Gamma})$ which are functorial in M.

We now define a Z-linear map $e:\Lambda\otimes\hat{\Gamma} \to \Gamma$ by $e(t^n\otimes f) = t^n f(t^n)$ for all f in $\hat{\Gamma}$ and n in Z. If we give $\Lambda\otimes\hat{\Gamma}$ the diagonal left Γ-module structure by $\theta t^s(t^r\otimes f) = t^{r+s}\otimes\theta t^s f$ for all f in $\hat{\Gamma}$, θ in Θ and r, s in Z then e is Γ-linear. We may use e to define cross products in cohomology. Let A_* be a Λ-chain complex and B_* a Γ-chain complex and give the tensor product $A_*\otimes B_*$ the total grading and differential, and the diagonal Γ-structure. Then there are cross products from $H^p(A_*;\Lambda)\otimes H^q(B_*;\hat{\Gamma})$ to $H^{p+q}(A_*\otimes B_*;\Gamma)$. If we take A_* to be a projective resolution of the Λ-module L and B_* to be a projective resolution of the Γ-module M we obtain cross products from $Ext^p_\Lambda(L,\Lambda)\otimes Ext^q_\Gamma(M,\hat{\Gamma})$ to $Ext^{p+q}_\Gamma(L\otimes M,\Gamma)$. In particular if $L = Z$ we have $Z\otimes M = M$ as Γ-modules and $Ext^1_\Lambda(Z,\Lambda) = Z$, and so we obtain maps from $Ext^q_\Gamma(M,\hat{\Gamma})$ to $Ext^{q+1}_\Gamma(M,\Gamma)$ which are functorial in M. Similarly if A_* is the chain complex $0 \to \Lambda \to \Lambda \to 0$ (concentrated in degrees 0 and 1) then there is a chain homotopy equivalence of $A_*\otimes B_*$ with B_*, and we obtain maps from $H^q(B_*;\hat{\Gamma})$ to $H^{q+1}(B_*;\Gamma)$ which are functorial in B_*.

Suppose now that B_* is a free Γ-chain complex such that $B_j = 0$ for $j < 0$ and $H_0(B_*) \cong Z$. Then the quotient map from B_0 onto $H_0(B_*)$ factors through Λ, and there is a chain homomorphism ε_* from B_* to the above complex A_*. Let η be the class in $H^1(B_*;\Lambda) = H^1(Hom_\Gamma(B_*,\Lambda))$ represented by $\varepsilon_1:B_1 \to \Lambda$. Then η is the image of a generator of $H^1(A_*;\Lambda)$ $(= Ext^1_\Lambda(Z,\Lambda)) \cong Z$. Now we may also obtain η as the image of a generator of $H^0(\ddot{B}_*;Z) = \{\lambda \text{ in } Hom_\Theta(B_0,Z) \mid \lambda\partial_1 = 0\} \cong Z$ under the analogous pairing from $Ext^1_\Lambda(Z,\Lambda)\otimes H^0(\ddot{B}_*;Z)$ to $H^1(B_*;\Lambda)$ of [Ba 1980']. Since $Ext^1_\Lambda(Z,\Lambda) = Ext^1_\Lambda(Z,\Lambda)\otimes H^0(\ddot{B}_*;Z) \cong Z$, our pairing from $Ext^1_\Lambda(Z,\Lambda)\otimes H^q(B_*;\hat{\Gamma})$ to $H^{q+1}(B_*;\Gamma)$ factors through $H^1(B_*;\Lambda)\otimes H^q(B_*;\hat{\Gamma})$ and we may identify our degree raising maps from $H^q(B_*;\hat{\Gamma})$ to $H^{q+1}(B_*:\Gamma)$ as those given by cup product with η.

PD-fibrations

If the homotopy fibre of a map from a PD_m-complex to a PD_n-complex is finitely dominated, then it is itself a PD_{m-n}-complex. (For a nice proof in the case when all the spaces are homotopy equivalent to finite complexes see [Go 1979]). We are interested in obtaining such a result for maps from PD_4^+-complexes to S^1, under weaker, purely algebraic hypotheses.

Let E be a connected PD_4^+-complex and let $G = \pi_1(E)$. Suppose that the map $f:E \to S^1$ induces an epimorphism $f_*:G \to Z$ with kernel H. Then we may identify the homotopy fibre of f with E', the covering space of E associated to the subgroup H, and we may recover E up to homotopy type as the mapping torus of a generator $\tau:E' \to E'$ of the covering group. If E' is a PD_3^+-complex then it is finitely dominated [Br 1972] and so H must be almost finitely presentable [Br 1975]. Moreover $H_*(E';Z)$ is then finitely generated and so $\chi(E) = 0$. We shall assume henceforth that these two conditions hold.

Let C_* be the equivariant cellular chain complex of the universal covering space $\tilde{E}$. Then there are Poincaré duality isomorphisms from $H^p(C_*;\Gamma)$ to $H_{4-p}(C_*)$ given by cap product with a generator $[E]$ of $H_4(C_*\otimes_\Gamma Z) \cong Z$. Since G is infinite $H^0(C_*;\Gamma) = 0$, and since $\tilde{E}$ is 1-connected $H^3(C_*;\Gamma) = 0$ also. When H is finite $\tilde{E}$ is homotopy equivalent to S^3 (as in Theorem 1 of Chapter 4) and so $E' = \tilde{E}/H$ is a PD_3^+-complex by [Wa 1967]. So henceforth we shall assume that H is infinite, and thus G has one end. Therefore we have $H^1(C_*;\Gamma) = 0$ also, so the only nonzero homology modules are $H_0(C_*) = Z$ and $\Pi = H_2(C_*) = \pi_2(E)$.

Since $H_1(C_*\otimes_\Gamma\Lambda) = H_1(E';Z) = H/H'$ is finitely generated as an abelian group, $Hom_\Gamma(H_1(C_*\otimes_\Gamma\Lambda),\Lambda) = 0$. An elementary computation then shows that $H^1(C_*;\Lambda)$ is infinite cyclic, and generated by the class η introduced in the previous section. By Poincaré duality for E with coefficients Λ the abelian group $H_3(E';Z) = H_3(C_*\otimes_\Gamma\Lambda)$ is infinite cyclic, generated by the class $[E'] = \eta\cap[E]$. Thus $H_i(E';Z) = H_i(\ddot{C}_*\otimes_\Theta Z) = H_i(C_*\otimes_\Gamma\Lambda)$ is finitely generated over Z and so is a torsion Λ-module, except perhaps when $i = 2$. On extending coefficients to $Q(t)$, the field of fractions of Λ, we may conclude that $H_2(C_*\otimes_\Gamma\Lambda)$ has rank $\chi(E) = 0$, and so is also a torsion Λ-module. Poincaré duality and the Universal Coefficient spectral sequence then imply that $H_2(C_*\otimes_\Gamma\Lambda) \cong Ext^1_\Lambda(H_1(C_*\otimes_\Gamma\Lambda),\Lambda)$, and so is finitely generated and torsion free over Z. Therefore E' satisfies Poincaré duality with coefficients Z, i.e. cap product with $[E']$ maps $H^p(\ddot{C}_*;Z)$ isomorphically to $H_{3-p}(\ddot{C}_*\otimes_\Theta Z) = H_{3-p}(C_*\otimes_\Gamma\Lambda)$ [Ba 1980'].

By standard properties of cap and cup products, to show that E' satisfies Poincaré duality with coefficients Θ, i.e. that cap product with $[E']$ gives isomorphisms from $H^p(\ddot{C}_*;\Theta)$ to $H_{3-p}(C_*)$ it shall suffice to show that the map η_p from $H^p(\ddot{C}_*;\Theta) = H^p(C_*;\hat{\Gamma})$ to $H^{p+1}(C_*;\Gamma)$ given by cup product with η is an isomorphism, for all p. (Cf. [Ba 1980']). There is nothing to prove when $p = 0$ or $p > 4$, for both modules are then 0. It remains to show that $H^p(\ddot{C}_*;\Theta) = 0$ if $p = 2$ or 4, and that η_1 and η_3 are isomorphisms.

It is clear that $H_p(\ddot{C}_*) = \ddot{H}_p$, where $H_p = H_p(C_*)$. The cohomology modules $H^p(\ddot{C}_*;\Theta)$ and $H^p(C_*;\Gamma)$ may be "computed" via Universal Coefficient spectral sequences, whose E_2^{p*} columns are nonzero only for $p = 0$ or 2. In particular, $H^1(\ddot{C}_*;\Phi) \cong Ext^1_\Phi(Z,\Phi)$, and if $Hom_\Gamma(\Pi,\Gamma) = 0$ then $H^2(C_*;\Gamma) \cong Ext^2_\Gamma(Z,\Gamma)$. Cup product with η determines maps between these spectral sequences which are compatible with the corresponding maps between the cohomology modules.

The E_2^{0*} terms of these spectral sequences involve only the cohomology of the groups G and H, and the maps between them may be identified with the maps arising in the LHS spectral sequence for G as an extension of Z by H. In particular since H is finitely generated $H^1(H;\Gamma) = \Lambda \otimes H^1(H;\Theta)$ and so $H^1(G/H;H^1(H;\Gamma)) = H^1(H;\Theta) = Ext^1_\Theta(Z,\Theta)$. If moreover H is finitely presentable then $H^2(H;\Gamma) = \Lambda \otimes H^2(H;\Theta)$, so $H^0(G/H;H^2(H;\Gamma)) = 0$. It then follows that the map from $Ext^1_\Theta(Z,\Theta)$ to $Ext^2_\Gamma(Z,\Gamma)$ is an isomorphism, and so η_1 is an isomorphism if $Hom_\Gamma(\Pi,\Gamma) = 0$.

At this point it becomes convenient to impose further restrictions on the group G and subgroup H.

Centre not contained in H

We shall consider first the case which is relevant to the study of twist spun knots. The appeal to smoothing theory in the next theorem in order to guarantee that $\ddot{C}_*$ is 3-dimensional seems rather artificial, but note that the theorem of Strebel establishing such a result when M is an aspherical PD_n-complex is by no means trivial [St 1977].

Theorem 3 *Let M be a closed orientable 4-manifold with $\chi(M) = 0$ and suppose that the map $f:M \to S^1$ induces an epimorphism f_* from $G = \pi_1(M)$ to Z such that ζG is not contained in $H = ker(f_*)$. Then the homotopy fibre M' of f is a PD_3^+-complex.*

Proof Let t be a central element of G which is not in H. Then the subgroup of G generated by t and H has finite index, and is isomorphic to $H \times Z$. Thus by passing to a finite cover we may assume that $G = H \times Z$. (It follows immediately that H is finitely presentable). Since M' is an open 4-manifold it is smoothable [Qu 1984], and so is homotopy equivalent to a 3-dimensional cell complex. Therefore $\ddot{C}_*$ is chain homotopy equivalent (over Θ) to a complex which is concentrated in degrees 0, 1, 2 and 3. In particular, $H^4(\ddot{C}_*;\Theta) = 0$.

We may localize the ring Γ with respect to the central multiplicative system S generated by $t-1$. Arguing as in Chapter 3, we find that Π_S is a stably free Γ_S-module of rank $\chi(M)$, and so is 0. Moreover $Hom_\Gamma(\Pi,\Gamma)$ embeds in $Hom_\Gamma(\Pi,\Gamma_S) = Hom_\Gamma(\Pi_S,\Gamma_S) = 0$ and so is also 0. (It follows that $\Pi = Ext^2_\Gamma(Z,\Gamma) = H^2(G;\Gamma)$). Hence η_1 is an isomorphism.

To handle the remaining entries, we shall relate Γ and $\hat{\Gamma}$ by an exact sequence whose other terms are (partially) divisible. (Cf. [Ba 1980]). Here we shall identify $\hat{\Gamma}$ with $\Theta[[t,t^{-1}]]$ via the map j introduced above. Define a Γ-linear map T from Γ_S to $\hat{\Gamma}$ by $T(\gamma/p(t)) = \gamma(Lexp_0(1/p(t)) - Lexp_\infty(1/p(t)))$ for all γ in Γ and $p(t)$ in S, where $Lexp_z(f(t))$ means the Laurent expansion of f at z. (Since the denominators $p(t)$ may all be assumed to be powers of $t-1$, it is easy to to make these expansions explicit. In particular $T(\gamma/(t-1)) = -\gamma\Sigma t^n$). Then $T(\Gamma_S)$ is the S-torsion Γ-submodule of $\hat{\Gamma}$, so $\Omega = Coker(T)$ is S-torsion free. It is easily verified that $(t-1)\hat{\Gamma} = \hat{\Gamma}$, and hence that $(t-1)\Omega = \Omega$. Therefore $t-1$ acts invertibly on Ω and so $\Omega_S = \Omega$. Thus there is a 4-term exact sequence of Γ-modules

$$0 \to \Gamma \to \Gamma_S \to \hat{\Gamma} \to \Omega \to 0$$

in which two of the terms are S-divisible. (Note that neither Γ nor $\hat{\Gamma}$ are Γ_S-modules). The localized complex C_{*S} is contractible, since $Z_S = \Pi_S = 0$. Therefore $H^*(C_*;\Gamma_S) = H^*(C_{*S};\Gamma_S) = 0$ and $H^*(C_*;\Omega) = H^*(C_{*S};\Omega_S) = 0$ and so $H^n(C_*;\hat{\Gamma})$ and $H^{n+1}(C_*;\Gamma)$ are isomorphic, for all n. In particular $H^2(\ddot{C}_*;\Theta) = H^2(C_*;\hat{\Gamma}) = H^3(C_*;\Gamma) = 0$ and $H^3(\ddot{C}_*;\Theta) = H^4(C_*;\Gamma) \cong Z$. As $\ddot{C}_*$ is chain homotopy equivalent to a 3-dimensional complex, the map from $H^3(\ddot{C}_*;\Theta)$ to $H^3(\ddot{C}_*;Z)$ induced by the augmentation of Θ onto Z is surjective. As remarked earlier, cap product with $[M']$ maps $H^3(\ddot{C}_*;Z)$ isomorphically to $H_0(\ddot{C}_*\otimes_\Theta Z) = Z$ [Ba 1980']. Since the composed map from $H^3(\ddot{C}_*;\Theta) = Z$ to $H_0(\ddot{C}_*) = Z$ may also be viewed as cap product with $[M']$, we conclude that the latter map is an isomorphism. Thus M' satisfies Poincaré duality with coefficients Θ. Since $\pi_1(M') = H$ is finitely presentable, M' is then finitely dominated [Br 1972], and so is a PD_3^+-complex. □

Corollary 1 *Let K be a 2-knot such that $\zeta\pi K$ is not contained in π'. Then the infinite cyclic cover M' of $M(K)$ is a PD_3^+-complex.* □

If π' has one end then M' is aspherical. (Cf. also Theorem 4 of Chapter 4). It remains an open question as to whether a PD_3^+-complex X such that $\pi_1(X)$ has infinitely many ends must be a connected sum [Wa 1967]. If $\pi_1(X)$ is not a generalized free product with amalgamation over a proper finite subgroup (e.g. if it is torsion free or if it is a 3-manifold group) then X is homotopy equivalent to a connected sum of PD_3^+-complexes, each of which is either aspherical, elliptical (finitely covered by a PD_3-complex homotopy equivalent to S^3) or $S^1\times S^2$ [Tu 1981]. We may ask whether any 2-knot with such a group factors in a related way.

Corollary 2 *Suppose that H is torsion free and that its indecomposable factors are fundamental groups of closed orientable 3-manifolds which are Haken, hyperbolic or Seifert fibred. Then M is homotopy equivalent to a PL 4-manifold which fibres over S^1.*

Proof By [He 1977] there is a homotopy equivalence $h:N \to M'$ from a closed orientable 3-manifold N whose factors are Haken, hyperbolic or Seifert fibred. Let $t:M' \to M'$ be a generator of the covering group of M' over M. As in Theorem 9 of Chapter 2 there is a self homeomorphism g of N which is homotopic to $h^{-1}th$. The conjugating map h gives rise to a homotopy equivalence from the mapping torus of g to M. □

If K is a classical knot and $r > 5$ then the commutator subgroup of $\pi\tau_r K$ is as in the corollary, and $M(\tau_r K)$ fibres over S^1.

The next result improves upon Corollary 3 of Theorem 3 of Chapter 3.

Corollary 3 *Let M be a closed orientable 4-manifold with $\chi(M) = 0$ and free abelian fundamental group. Then M is homotopy equivalent to $S^3 \times S^1$, $S^2 \times S^1 \times S^1$, or $S^1 \times S^1 \times S^1 \times S^1$.* □

It can be shown that any such manifold M is homeomorphic to one of these standard examples [Kw 1986].

Groups of finite geometric dimension 2

In this section we shall assume that G has finite geometric dimension 2, deficiency 1 and one end. For any (left) Γ-module N let $e^q N$ be the left Γ-module which has the same underlying abelian group as the right Γ-module $Ext^q_\Gamma(N,\Gamma)$ with the conjugate left Γ-module structure. Since $c.d.G = 2$, the ring Γ has global dimension 3 and so $e^q N = 0$ for $q > 3$. Moreover if $\Gamma^g \to \Gamma \to Z \to 0$ is a partial projective resolution of the augmentation Γ-module then the kernel at the left, L say, is projective. Since G has a finite 2-complex as an Eilenberg-Mac Lane space and deficiency 1, we may assume that L is free of rank $g-1$. Since G has one end, $e^0 Z = e^1 Z = e^3 Z = 0$ and so transposing this resolution gives a resolution $0 \to \Gamma \to \Gamma^g \to \Gamma^{g-1} \to e^2 Z \to 0$ for $e^2 Z$. We then see that $e^0 e^2 Z = e^1 e^2 Z = e^3 e^2 Z = 0$ and $e^2 e^2 Z = Z$. In particular, $e^2 Z$ is nonzero, and by the functoriality of $e^2(-)$ we have $Aut(e^2 Z) = Aut(Z) = \pm 1$.

Theorem 4 *Let M be a closed orientable 4-manifold with $\chi(M) = 0$ and such that $G = \pi_1(M)$ has finite geometric dimension 2, deficiency 1 and one end. Then $\pi_2(M) \cong e^2Z$, and the isomorphism is unique up to sign.*

Proof Let $C_* = 0 \to C_4 \to C_3 \to C_2 \to C_1 \to C_0 \to 0$ be the cellular chain complex of the universal covering space $\widetilde{M}$. Then C_* is a complex of finitely generated free left Γ-modules whose homology is $H_* = H_*(M;\Gamma) = H_*(\widetilde{M};Z)$. By Poincaré duality and the Universal Coefficient spectral sequence $H_j = 0$ unless $j = 0$ or 2, while $e^1H_2 = e^3H_2 = 0$, $e^2H_2 = Z$ and there is an exact sequence of left Γ-modules $0 \to e^2Z \to H_2 \to e^0H_2 \to 0$. Applying the functor $e^*(-)$ to this exact sequence, and writing P for e^0H_2, we obtain a twelve term exact sequence $0 \to e^0P \to \cdots \to e^3e^3Z \to 0$. Using the above observations, we find that $e^1P = 0$, and that there is an exact sequence $0 \to e^2P \to e^2H_2 \to e^2e^2Z \to e^3P \to 0$. But $e^3P = 0$ since P is a submodule of a free module and Γ has global dimension 3. Since $e^2H_2 = e^2e^2Z = Z$, it follows that $e^2P = 0$ also. Hence P is projective, and so $H_2 = P\oplus e^2Z$. Therefore we have exact sequences

$$0 \to C_4 \to C_3 \to Z_2 \to H_2 = P\oplus e^2Z \to 0 \tag{1}$$

and

$$0 \to Z_2 \to C_2 \to C_1 \to C_0 \to Z \to 0 \tag{2}$$

where Z_2 is the module of 2-cycles, which is projective, since (from (5)) it is a third syzygy. Now $P\oplus e^2Z$ also has a resolution of the form

$$0 \to \Gamma \to \Gamma^g \to P\oplus\Gamma^{g-1} \to P\oplus e^2Z \to 0. \tag{3}$$

Applying Schanuel's lemma to (1) and (3) and to (2) and the resolution of Z given above, we find that $C_4\oplus\Gamma^g\oplus Z_2 = \Gamma\oplus C_3\oplus P\oplus\Gamma^{g-1}$ and $Z_2\oplus\Gamma^{g-1}\oplus C_1\oplus\Gamma = C_2\oplus\Gamma^g\oplus C_0$ and therefore $C_4\oplus\Gamma^g\oplus C_2\oplus\Gamma^g\oplus C_0 = \Gamma\oplus C_3\oplus\Gamma^{g-1}\oplus P\oplus\Gamma^{g-1}\oplus C_1\oplus\Gamma$. But $\chi(M) = 0$, so $C_4\oplus C_2\oplus C_0 = C_3\oplus C_1$. Thus there is a finitely generated free Γ-module $F = C_3\oplus C_1\oplus\Gamma^{2g}$ such that

$F = F\oplus P$. This implies that $P = 0$, by Kaplansky's Lemma, so $H_2 = e^2Z$ and the first assertion follows from the Hurewicz Theorem. The isomorphism is unique up to sign since $Aut(e^2Z) = (\pm 1)$. □

Is this theorem still true under the formally weaker assumption that G has cohomological dimension 2? It is not yet known whether each such group has geometric dimension 2. (Cf. Theorem 6 of Chapter 2). Whether such a knot group has one end is related to the question of Kervaire as to whether a nontrivial free product with abelianization Z can have weight 1 [Ke 1965].

Corollary 1 *The cellular chain complex of $\widetilde{M}(K)$ is determined up to chain homotopy equivalence over Γ by G.*

Proof There are exact sequences $0 \to B_1 \to C_1 \to C_0 \to Z \to 0$ and $0 \to C_4 \to C_3 \to Z_2 \to \Pi \to 0$. Schanuel's lemma implies that B_1 is projective, since $c.d.G = 2$. Therefore $C_2 \cong Z_2\oplus B_1$ and so Z_2 is also projective. Thus C_* is the direct sum of a projective resolution of Z and a projective resolution of $\Pi \cong e^2Z$ with degree shifted by 2. □

In general there may be an obstruction to realizing a chain homotopy equivalence between two such chain complexes by a map of spaces (cf. [Ba 1986]).

Corollary 2 *Let K and K_1 be 2-knots with group G. Then there is a 3-connected map $f_o:M(K_1)-int\,D^4 \to M(K)$. If f_o extends to a map $f:M(K_1) \to M(K)$, then f is a homotopy equivalence.*

Proof In each case $H^3(G;\pi_2(M))$ is trivial, since $c.d.G = 2$, and so the first k-invariant is 0. The existence of a map f_o which induces isomorphisms on π_1 and π_2 now follows as in Theorem 1. Since $H_3(\widetilde{M};Z) = H_3(M;\Gamma) = 0$, $\pi_3(M) = \Gamma_W(\pi_2(M))$ (where $\Gamma_W(-)$ is here the quadratic functor of Whitehead [Wh 1950]), and therefore f_o induces an epimorphism on π_3, and so is

3-connected. If f extends f_o then it must induce isomorphisms on the homology of the universal covering spaces, since $H_j(M(K_1;\Gamma)) = H_j(M(K);\Gamma)) = 0$ for $j \geqslant 3$. Therefore f is a homotopy equivalence. □

Thus the major task in determining the homotopy type of $M(K)$ is to decide when f_o extends. When πK is such a knot group and π' is finitely generated then it must be free [B: Corollary 8.6], and π then determines the homotopy type of M, as we shall show next.

Free kernel

We shall now assume that G is an extension of Z by a finitely generated free normal subgroup, and shall adapt our earlier notation without further comment.

Theorem 5 *Let M be a closed orientable 4-manifold with $\chi(M) = 0$ and suppose that the map $f:M \rightarrow S^1$ induces an epimorphism f_* from $G = \pi_1(M)$ to Z with kernel H a free group of rank r. Then M is simple homotopy equivalent to a PL 4-manifold N which fibres over S^1 with fibre $\#^r(S^1 \times S^2)$ and which is determined up to PL homeomorphism by its fundamental group G.*

Proof If $r = 0$ then the argument of Theorem 1 shows that M is homotopy equivalent to $S^3 \times S^1$. Therefore we shall assume that $r > 0$. Since G then has a nontrivial finitely generated free normal subgroup of infinite index, it has finite geometric dimension 2, deficiency 1 and one end. Therefore $\Pi = e^2 Z$, by Theorem 4, and so $e^q \Pi = Z$ if $q = 2$ and is 0 otherwise. In particular, $Hom_\Gamma(\Pi,\Gamma) = 0$ and so η_1 is an isomorphism. Moreover, $e^2 Z = Ext^2_\Gamma(Z,\Gamma)$ is isomorphic to $Ext^1_\Theta(Z,\Theta)$ as a (left) Θ-module, so we also have $Ext^q_\Theta(\ddot{\Pi},\Theta) = Z$ if $q = 1$ and is 0 otherwise. From the spectral sequence for $H^*(\ddot{C}_*;\Theta)$ we then find that $H^2(\ddot{C}_*;\Theta) = H^4(\ddot{C}_*;\Theta) = 0$ and $H^3(\ddot{C}_*;\Theta) \cong Z$. Therefore $\ddot{C}_*$ is chain homotopy equivalent (over Θ) to a complex which is concentrated in degrees 0, 1, 2 and 3. It then follows that the map from $H^3(\ddot{C}_*;\Theta)$ to $H^3(\ddot{C}_*;Z)$ induced by the augmentation of Θ

onto Z is onto, and so is an isomorphism, as each module is isomorphic to Z. Similarly the map from $H^4(C_*;\Gamma)$ to $H^4(C_*;\Lambda)$ induced by the projection of Γ onto Λ is also an isomorphism. There is a commutative square

$$\begin{array}{ccc} H^3(\ddot{C}_*;\Theta) & \longrightarrow & H^4(C_*;\Gamma) \\ \downarrow & & \downarrow \\ H^3(\ddot{C}_*;Z) & \longrightarrow & H^4(C_*;\Lambda) \end{array}$$

in which the vertical maps are the isomorphisms just described and the horizontal maps are given by cup product with η. As remarked earlier, $H_*(C_*;\Lambda)$ is finitely generated over Z and so the lower map is an isomorphism [Ba 1980']. Therefore the upper map is also an isomorphism.

Since H is free, $K_0(H) = Wh(H) = 0$ [Wa 1978] and so M' is a finite simple PD_3^+-complex. Now such a complex is determined up to homotopy type by its fundamental group and a class τ in $H_3(\pi_1(M');Z)$ [He 1977]. Therefore M' is homotopy equivalent to $\#^r(S^1\times S^2)$. As every self homotopy equivalence of $\#^r(S^1\times S^2)$ is homotopic to a PL homeomorphism (which is unique up to isotopy [La 1974]), M is homotopy equivalent to the mapping torus N of such a homeomorphism. Finally any such homotopy equivalence is simple, as G is in Waldhausen's class Cl and so $Wh(G) = 0$ by Theorem 19.4 of [Wa 1978]. □

If K is the Artin spin of a classical fibred knot then $M(K)$ fibres over S^1 with such a fibre. However not all such fibred 2-knots are obtained in this way. (For instance, the Alexander polynomial need not be symmetric [AY 1981]). A problem of interest here which was raised by Neuwirth [N: Problem P] and is still open is: which groups are the groups of classical fibred knots? Apart from the trefoil knot group and the group of the figure eight knot there are just two groups G with G' free of rank 2 and $G/G' = Z$ [Ra 1960]; as they each have weight 1 and deficiency 1 they are 2-knot groups. So also is the group with presentation $\langle x,y \mid x^2y^2x^2 = y\rangle$ which is the group of a fibred knot in the Brieskorn homology 3-sphere $M(2,3,11)$, but which is not the group of any classical knot [Ra 1983].

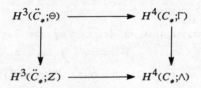

Quasifibres and minimal Seifert hypersurfaces

Let M be a closed orientable 4-manifold and $f:M \to S^1$ a map which induces an epimorphism f_* from $G = \pi_1(M)$ to Z and which is transverse over p in S^1. Then $\hat{V} = f^{-1}(p)$ is a codimension 1 submanifold with a product neighbourhood $N \simeq \hat{V}\times[-1,1]$. Let $W = M-\hat{V}\times(-1,1)$ and $\partial_\pm W = \hat{V}\times\{\pm 1\}$. We shall say that $\hat{V}$ is a *quasifibre* for f if the inclusions $j_\pm$ of $\hat{V}$ into W as $\partial_\pm W$ each induce monomorphisms on fundamental groups. By van Kampen's theorem G is then an HNN extension with base $\pi_1(W)$ and associated subgroups $j_{\pm *}(\pi_1(\hat{V}))$. If f is the projection of a fibre bundle then every fibre is a quasifibre, and $H = ker\, f_*$ is finitely generated. The next result provides a partial converse to this observation.

Theorem 6 *Let M be a closed orientable 4-manifold with $\chi(M) = 0$ and suppose that the map $f:M \to S^1$ induces an epimorphism f_* from $G = \pi_1(M)$ to Z with finitely generated kernel H. If f has a quasi-fibre $\hat{V}$ then W (as above) is an h-cobordism. Hence M' is homotopy equivalent to $\hat{V}$.*

Proof Note first that as H is finitely generated the monomorphisms $j_{\pm *}$ must in fact be isomorphisms, and therefore $H^s(W,\partial_\pm W;\Gamma) = H_s(W,\partial_\pm W;\Gamma) = 0$ for $s = 0$ or 1. By Poincaré-Lefshetz duality, these modules are also 0 for $s = 3$ or 4. Therefore $H_2(W,\partial_\pm W;\Gamma)$ is a stably free Γ-module, by [W: Lemma 2.3]. As in Theorem 3 of Chapter 3, we may compute its rank as $\chi(W,\partial_\pm W) = \chi(W)-\chi(\hat{V})$. As $\hat{V}$ is a closed 3-manifold, $\chi(\hat{V}) = 0$, and as $\chi(M) = 0$ it then follows that $\chi(W) = 0$, and so the module is 0, by Kaplansky's Lemma. Therefore each of the inclusions $j_\pm$ is a homotopy equivalence, and W is an h-cobordism. The last assertion follows easily. □

We shall say that a Seifert hypersurface V for a 2-knot K is *minimal* if $\hat{V} = V\cup D^3$ is a quasifibre for some map $f:M(K) \to S^1$.

Corollary *Let K be a 2-knot such that π' is finitely generated, and* *which has a minimal Seifert hypersurface $\hat{V}$. If every self homotopy equivalence of $\hat{V}$ is homotopic to a homeomorphism then $M(K)$ is homotopy equivalent to $M(K_1)$, where K_1 is a fibred 2-knot with closed fibre $\hat{V}$.*

Proof Let j_+^{-1} be a map from M' to $\hat{V}$ which is a homotopy inverse to the homotopy equivalence j_+, and let θ be a self homeomorphism of $\hat{V}$ homotopic to $j_+^{-1}j_-$. Then $j_+\theta j_+^{-1}$ is homotopic to a generator of the covering translations of M', and so the mapping torus of θ is homotopy equivalent to M. As in Theorem 9 of Chapter 2 surgery on this mapping torus gives such a knot K_1. □

If a Seifert hypersurface V for a 2-knot has fundamental group Z, then V is minimal, as can be seen from the Mayer-Vietoris sequence for $H_*(M;\Lambda)$ (cf. [H: page 14]). Examples 10 and 11 of [Fo 1962] are ribbon 2-knots with such minimal Seifert hypersurfaces (homeomorphic to $S^1 \times S^2 - int\, D^3$) but for which π' is not finitely generated.

In Chapter 8 we shall show that if the 3-dimensional Poincaré conjecture is true then there are 2-knots with π' finite which have no minimal Seifert surface. For some interesting ideas on avoiding this problem see [Gu 1979]. (Note however that the main theorem of [Gu 1978] is contradicted by [Yo 1988], where it is shown that certain HNN extensions with base a torus knot group and associated subgroups infinite cyclic are the groups of ribbon 2-knots but cannot be expressed as HNN extensions with free base).

Further remarks

We may ask what determines the homotopy type of $M(K)$ for more general 2-knots K. (In view of Theorem 1 we may assume that π' is infinite). A good candidate for an algebraic invariant is the *algebraic 3-type* (π,Π,k), where $\Pi = \pi_2(M)$, considered as a $Z[\pi]$-module, and k is the first k-invariant in $H^3(\pi;\Pi)$ [MW 1950]. The algebraic 3-type of M' is then $(\pi',\ddot{\Pi},\ddot{k})$, where $\ddot{k}$ is the image of k in $H^3(\pi;\ddot{\Pi})$ and $\ddot{}$ is the forgetful functor from $Z[\pi]$-modules to $Z[\pi']$-modules.

Suppose that K_1 is another 2-knot such that there is an isomorphism (α,β) from the algebraic 3-type of M to that of $M_1 = M(K_1)$. (Thus α is a π_1-isomorphism, β is a π_2-isomorphism which respects the π_1-actions (via α) and the k-invariants correspond). This induces an isomorphism between the algebraic 3-types of the infinite cyclic covering spaces M' and M_1'. As these are homotopy equivalent to 3-complexes, this isomorphism is induced by a map $f:M' \to M_1'$. If π has one end then $\pi_3 = \Gamma_W(\Pi)$ and so f is in fact a homotopy equivalence. Let t and t_1 be generators of the infinite cyclic covering groups which correspond under α. Then the composite $f^{-1}t_1^{-1}ft$ is a self homotopy equivalence of M' which fixes the algebraic 3-type. If this is homotopic to the identity then the mapping tori of t and t_1 are homotopy equivalent. Up to homotopy these are just M and M_1 respectively. Thus the problem may be reduced to determining the obstructions to constructing a homotopy from a self map of a 3-complex which fixes its algebraic 3-type to the identity.

The related problem of determing the homotopy type of the exterior of a 2-knot has been considered by Lomonaco, Plotnick and Suciu [Lo 1981, Pl 1983, PS 1985]. (Note that $\Pi = \pi_2(X)/\langle\partial\rangle$, where $\langle\partial\rangle$ is the π-submodule of $\pi_2(X)$ generated by a "longitudinal" 2-sphere in $\partial X \cong S^2\times S^1$). In each of the examples considered in [Pl 1983] either π' is finite or M is aspherical, and so they do not test the adequacy of the algebraic 3-type for the present problem. On the other hand the examples of [PS 1985] probably also show that in general M is not determined by π and Π alone.

Baues has developed algebraic classifications of 4-dimensional complexes and 4-dimensional Poincaré complexes [Ba 1986]. However it is not yet clear how they may be applied to our problem. For other recent work on the homotopy type of 4-manifolds (albeit with finite fundamental group), see [HK 1988].

Chapter 8 APPLYING SURGERY TO DETERMINE THE KNOT

There are several potential difficulties in attempting to apply surgery effectively to the study of 2-knots. Firstly, the 4-dimensional disk embedding theorem central to the theory has only been established over fundamental groups in a limited class. Secondly the surgery obstructions are notoriously difficult to compute. (Finally we might add that the extension of the theory to the PL or DIFF context remains a mystery).

The class of groups over which 4-dimensional TOP surgery sequences are exact includes all the finitely presentable groups with ascending series of the type considered in Chapter 6. If we assume that π is such a 2-knot group and moreover that it has an abelian normal subgroup of positive rank then either $M(K)$ is aspherical (and π is virtually poly-Z) or π' is finite or $\pi = \Phi$. When π is a torsion free virtually poly-Z group the surgery obstructions vanish, and when π is poly-Z it determines the knot up to Gluck reconstruction and change of orientations. We shall consider in detail the effect of Gluck reconstruction on such knots.

When π' is finite the difficulties in classifying such knots lie in determining Whitehead torsions, the surgery obstruction groups $L_5^s(\pi)$ and their action on the structure sets. However some computations are feasible, and we can show that there are infinitely many 2-knots K such that $M(K)$ is simple homotopy equivalent to $M(\tau_2 T)$, where T is the trefoil knot. If the 3-dimensional Poincaré conjecture is true, then among these knots only $\tau_2 T$ has a minimal Seifert manifold, and the manifolds $M(K)$ for the other knots are counter examples to the 4-dimensional analogue of the Farrell fibration theorem.

In the case of the group Φ the Whitehead group is 0 and the Wall group acts trivially, and thus the homotopy type of $M(K)$ determines the exterior of the knot. (Thus the only difficulty in characterizing such knots is in constructing a homotopy equivalence from $M(K)$ to a standard model). The Whitehead group is also trivial whenever π' is free or is the group of an aspherical Seifert fibred 3-manifold. Moreover in most of these cases the Wall group acts trivially, and only the s-cobordism theorem is needed to complete the classification. We introduce the equivalence relation of "s-concordance" between 2-knots in order to sidestep this problem.

Preliminaries

If M is a closed n-manifold we shall let $S_{TOP}(M)$ denote the structure set of simple homotopy equivalences with target M, modulo s-cobordism. When $\pi_1(M)$ is in Freedman's class this agrees with the usual definition. We shall let $\theta_i(M)$ be the surgery obstruction map from $[S^{i-n}(M^+),G/TOP]$ to $L_i^s(\pi_1(M))$. As all the manifolds to be considered here are orientable, we may assume that all orientation characters are trivial, and suppress them from the notation.

If we fix an isomorphism $i_Z:Z \to L_5(Z)$ then we may define a homomorphism $I_G:G \to L_5^s(G)$ for any group G by $I_G(g) = g_*(i_Z(1))$, where $g_*:Z = L_5(Z) \to L_5^s(G)$ is induced by the map sending 1 in Z to g in G [We 1983]. This homomorphism clearly factors through the abelianization G/G', and $I_Z = i_Z$. Moreover I_G is natural in the sense that if $f:G \to H$ is a homomorphism then $L_5(f)I_G = I_H f$. Therefore if $\alpha:G \to Z$ induces an isomorphism on abelianization the homomorphism $\hat{I}_G = I_G\alpha^{-1}I_Z^{-1}$ is a canonical splitting for $L_5(\alpha)$.

Lemma 1 *Let K be a 2-knot with group π. Then*

(*i*) *the surgery obstruction map $\theta_4:[M(K),G/TOP] \to L_4^s(\pi)$ is injective;*

(*ii*) *the image of $L_5(Z)$ under $\hat{I}_\pi$ acts trivially on $S_{TOP}(M(K))$;*

(*iii*) *if $\chi(N) = 0$ then any 2-connected degree 1 map $f:N \to M(K)$ is a homotopy equivalence.*

Proof Since $[M,G/TOP] = Z$ and the composition with the projection onto $L_4(1)$ is given by the signature difference map (cf. [W: page 237]), (i) is clear.

Let P be the E_8 manifold [Fr 1982] and delete the interior of a submanifold homeomorphic to $D^3\times I$ to obtain P_o. There is a normal map $p:P_o \to D^3\times I$ which we may assume is the identity on the boundary (and in particular respects the corners of the boundary). Let U be a regular neighbourhood of a meridian for M and form the union of $(M-int\,U)\times I$ with $P_o\times S^1$, identifying $\partial U\times I = S^2\times S^1\times I$ with $S^2\times I\times S^1$ in $\partial P_o\times S^1$. We may also match together $id_{(M-int\ U)\times I}$ and $p\times id_{S^1}$ to obtain a normal cobordism Q from id_M to itself. The surgery obstruction of this normal cobordism generates the image of $\hat{I}_\pi$, which proves (ii).

If $f:N \to M$ is a 2-connected degree 1 map then its failure to be an homotopy equivalence is measured by the homology kernel $K_2(f)$, which is a stably free $Z[\pi]$-module [W: Proposition 2.3]. As in Chapter 3 we see that this module has rank $\chi(N)-\chi(M)$, and so (iii) follows from Kaplansky's Lemma. □

Lemma 2 *If the abelianization homomorphism induces an isomorphism from $L_5^s(\pi)$ to $L_5(Z)$ then simple homotopy equivalent closed orientable 4-manifolds with fundamental group π are s-cobordant.*

Proof By part (i) of Lemma 1 any such simple homotopy equivalence $f:N \to M$ is normally cobordant to id_M, and by part (ii) and our assumption on $L_5^s(\pi)$ we may find a normal cobordism with obstruction 0. □

At present there is not yet a "plumbing theorem" over arbitrary groups: we do not know whether the whole of $L_5^s(\pi)$ acts on $S_{TOP}(M)$, except when π is in Freedman's class, or when Lemma 2 applies. (This is so for instance if π is square root closed accessible [Ca 1973]).

We shall say that two 2-knots K_0 and K_1 are *s-concordant* if there is a concordance $K:S^2\times I \to S^4\times I$ whose exterior is an s-cobordism (*rel* ∂) from $X(K_0)$ to $X(K_1)$. Surgery on K then gives an s-cobordism from $M(K_0)$ to $M(K_1)$ in which the meridians for K_0 and K_1 are conjugate. Conversely if $M(K)$ and $M(K_1)$ are s-cobordant via an s-cobordism of this kind then K_1 is s-concordant to K or K^*. In particular, if K is reflexive then K and K_1 are s-concordant.

The aspherical case

We have seen that whenever the group π of a 2-knot K contains a sufficiently large abelian normal subgroup then $M(K)$ is aspherical. This is notably the case when K is the q-twist spin of a prime knot for some $q \geqslant 3$ (excepting the 3-, 4- and 5-twist spins of the trefoil knot). In the latter case the universal cover $\widetilde{M}$ is homeomorphic to R^4. This is so whenever the group is simply connected at infinity. (Note however that there are aspherical closed 4-manifolds which are not covered by R^4 [Da 1983]).

Theorem 1 *Let K be a 2-knot whose group π has a nontrivial torsion free abelian normal subgroup A, and suppose that if A has rank 1 then π/A has one end. Then the universal cover $\widetilde{M}(K)$ is homeomorphic to R^4.*

Proof The assumptions imply that M is aspherical, and hence that $\widetilde{M}$ is a contractible open 4-manifold. Suppose that A has rank 2. Then the virtual homological dimension of π/A is at least $4-h.d.A \geqslant 2$, and so π/A cannot have 0 or 2 ends. If it had infinitely many ends then it would have a generalized free product structure, with amalgamation over a finite subgroup, and so π would have a corresponding structure with amalgamation over a finite extension of A. The building blocks of such a structure must have infinite index in π and so have homological dimension less than 4, by [B: Proposition 9.22], and the amalgamating subgroup would have homological dimension 2. A Mayer-Vietoris calculation would then give $H_4(\pi;Z) = 0$, contrary to π being a PD_4-group. Thus $e(\pi/A) = 1$ in this case also. If A has rank 4 then $\zeta\pi = Z$ and $e(\pi/\zeta\pi) = 1$. Thus in all cases π is 1-connected at ∞, by Theorems 1 and 2 of [Mi 1987]. Therefore $\widetilde{M}$ is also 1-connected at ∞ and so is homeomorphic to R^4 by [Fr 1982]. □

The cases when the rank of A is greater than 2 follow also from the next theorem, using nonsimply-connected surgery instead of the results of [Mi 1987]. The following lemma is adapted from [W: Theorem 15.B.1].

Lemma 3 [W] *Let M' be an aspherical $(n-1)$-manifold and let M be the mapping torus of a self homeomorphism of M'. Suppose that $Wh(\pi_1(M'))$ is 0. If the surgery obstruction maps $\theta_i(M')$ are isomorphisms for all (sufficiently large) i then so are the maps $\theta_i(M)$.*

Proof This is an application of the 5-lemma and periodicity as in pages 229-230 of [W]. □

The hypotheses of this lemma are satisfied if M' is an irreducible closed orientable 3-manifold and $\pi_1(M')$ is square root closed accessible [Ca 1973], or is virtually nilpotent [FH 1983], or if M' admits an effective S^1-action with orientable orbit space (excepting perhaps some cases when the orbit space is S^2) [St 1985], or if M' is hyperbolic [FJ 1988].

Theorem 2 *Let K be a 2-knot whose group π is torsion free and solvable, but is neither Z nor Φ. Then K is determined up to Gluck reconstruction by π together with a generator of $H_4(\pi;Z)$ and the strict weight orbit of a meridian.*

Proof Let K_1 be another 2-knot with such a group. Then $M(K)$ and $M(K_1)$ are aspherical, by Theorem 3 of Chapter 3. Therefore any isomorphism from $\pi(K_1)$ to $\pi(K)$ is induced by a homotopy equivalence from $M(K_1)$ to $M(K)$. Since π is torsion free and virtually poly-Z, $Wh(\pi) = 0$ by [FH 1981], and so any such homotopy equivalence must be simple. Moreover π' is virtually nilpotent, by the work of Chapter 6. Therefore the surgery obstruction map from $[S^{i-3}(K(\pi',1)^+),G/TOP]$ to $L_i(\pi')$ is an isomorphism for $i \geqslant 6$, by [FH 1983]. Lemma 3 then implies that the maps $\theta_4(M)$ and $\theta_5(M)$ are isomorphisms. Since M is aspherical orientations of M correspond to generators of $H_4(\pi;Z)$, and so the theorem now follows from the exactness of the surgery sequence and our earlier work. □

Theorem 2 applies in particular to the examples of Cappell and Shaneson. (For such knots the strict weight orbit is unique up to inversion, by Theorem 8 of Chapter 2). A similar argument shows that, when they exist, Cappell-Shaneson n-knots are determined (up to Gluck reconstruction and change of orientation) by their groups together with the condition that $\pi_i(X) = 0$ for $2 \leqslant i \leqslant (n+1)/2$.

Theorem 3 *Let K be a 2-knot with group π such that π' is almost* *finitely presentable, π'/π'' is infinite and $\zeta\pi'$ is nontrivial. Then K is s-concordant to a fibred knot with closed fibre an aspherical Seifert fibred 3-manifold, which is determined (among such fibred knots) up to Gluck reconstruction by π together with a generator of $H_4(\pi;Z)$ and the strict weight orbit of a meridian.*

Proof The manifold $M(K)$ is aspherical and the commutator subgroup π' is the fundamental group of an aspherical closed orientable Seifert fibred 3-manifold, N say, by Theorem 5 of Chapter 4 and Theorem 6 of Chapter 5. Now π' has a subgroup σ of finite index which is a central extension of the group of a closed surface by Z, and so is the amalgamation of $Z\times F(m)$ with $Z\times F(n)$ over Z^2 for suitable free groups $F(m)$ and $F(n)$. Therefore $Z[\sigma]$ and hence $Z[\pi']$ are regular coherent rings and so $Wh(\pi) = 0$, by the Mayer-Vietoris sequence of [Wa 1978]. Since π'/π'' is infinite and has nontrivial centre, N is sufficiently large and admits an effective S^1-action [Wa 1967], and the orbit surface has positive genus. Therefore the surgery obstruction maps θ_j over π are isomorphisms for j large, by [St 1985] and Lemma 3. Thus $M(K)$ is determined up to s-cobordism by π, and the rest of the theorem follows easily. □

This theorem applies to most branched twist spins of torus knots. (Cf. the remarks preceding Theorem 7 of Chapter 5). The Brieskorn manifold $M(p,q,r)$ admits an effective S^1-action with orientable orbit space, and the orbit space has genus $\geqslant 1$ if and only if p, q and r are not all pairwise relatively prime. In fact, the hypotheses of Lemma 3 hold for all aspherical Brieskorn manifolds, except perhaps for $M(2s,3,5)$, $M(2,3t,5)$ and $M(2,3,5u)$, where s, t and u are odd primes such that $(s,15) = (t,5) = (u,3) = 1$, by the results of [St 1985]. If $Wh = 0$, the 5-dimensional s-cobordism theorem and a strong form of the Novikov conjecture can be established for all orientable Poincaré duality groups then any 2-knot with such a group would also be determined up to Gluck reconstruction by its group together with a generator of $H_4(\pi;Z)$ and a strict weight orbit.

The argument of Cappell and Shaneson

Cappell and Shaneson showed that if none of the eigenvalues of the monodromy of a fibred n-knot K with $M(K)$ an $(S^1)^{n+1}$ bundle over S^1 are negative then K and K^* are distinct. Such knots are strongly $(-1)^{n+1}$-amphicheiral, since inversion in each fibre gives an involution of $M(K)$ fixing a circle. However when n is even such knots are not invertible, for the Alexander polynomial (i.e. the characteristic polynomial of the monodromy) then has odd degree and does not vanish at 1 or -1, and so cannot be symmetric. Thus for each such 2-knot there are 4 distinct knots with the same group.

Instead of repeating the argument of [CS 1976] verbatim, we shall adapt it to answer fully the corresponding question for the 2-knots with $\pi' = \Gamma_q$ for some odd q. (Notable among these is the 6-twist spin of the trefoil knot). Like the examples of Cappell and Shaneson, these knots are fibred with closed fibre a coset space of a 3-dimensional Lie group. (In fact in many cases the manifold $M(K)$ admits a 4-dimensional geometry. See Appendix A).

Let Nil be the subgroup of $SL(3,R)$ consisting of the upper triangular matrices $[r,s,t] = \begin{pmatrix} 1 & r & t \\ 0 & 1 & s \\ 0 & 0 & 1 \end{pmatrix}$ for r, s and t in R. The group Nil is a 3-dimensional nilpotent Lie group, with abelianization R^2 and centre $\zeta Nil = R$. As a space Nil is canonically homeomorphic to R^3. The elements $x = [1,0,0]$, $y = [0,1,0]$ and $z = [0,0,1/q]$ generate a discrete subgroup of Nil isomorphic to Γ_q, and the coset space $N_q = Nil/\Gamma_q$ is a closed 3-manifold. The action of ζNil on Nil induces a free action of $S^1 = \zeta Nil/\zeta\Gamma_q$ on N_q which is a circle bundle over the torus. We shall take $[0,0,0]$ as the base point for Nil, and its image as the base point for such coset spaces.

The abelianization homomorphism induces a natural map from $Aut_{Lie}(Nil)$ to $Aut_{Lie}(R^2) = GL(2,R)$, whose kernel is isomorphic to $Hom_{Lie}(Nil,\zeta Nil) = R^2$. From this it is not hard to see that the group $Aut_{Lie}(Nil)$ may be described as the set of ordered pairs $GL(2,R)\times R^2$, with $(A,\mu) = (\begin{pmatrix} a & c \\ b & d \end{pmatrix}, (e,f))$ acting via $(A,\mu)([r,s,t]) = [ar+cs,\ br+ds,$

$er+fs+(ad-bc)t+bcrs+½(abr(r+1)+cds(s+1))]$. The action is clearly orientation preserving, as its Jacobian is everywhere $(ad-bc)^2$. If (B,ν) is another automorphism, with $B = \begin{pmatrix} g & j \\ h & k \end{pmatrix}$, then the product of (A,μ) and (B,ν) is $(AB,\mu B+(\det A)\nu+½\eta(A,B))$, where $\eta(A,B)$ is the vector $(abg(1-g)+cdh(1-h)-2bcgh,\ abj(1-j)+cdk(1-k)-2bcjk)$. In particular the group is *not* a semidirect product of $GL(2,R)$ with R^2. For each q, the subset $GL(2,Z)\times(q^{-1}Z)^2$ is a subgroup, and an analogous argument shows that this is $Aut(\Gamma_q)$. Thus every automorphism of Γ_q extends to an automorphism of Nil and so induces a base point and orientation preserving self homeomorphism of N_q.

As we showed in Theorem 10 of Chapter 6, each meridianal automorphism of $\Gamma = \Gamma_1$ is conjugate in $Out(\Gamma)$ to one of the form $(D,(0,0))$ where D in $GL(2,Z)$ has characteristic polynomial X^2-X+1, X^2-3X+1, X^2+X-1 or X^2-X-1. The characteristic polynomial determines the conjugacy class. The latter two also represent the two classes of meridianal automorphisms of Γ_q with q odd and greater than 1.

If K is a 2-knot with $\pi K' = \Gamma_q$ then $M(K)$ is homeomorphic to the mapping torus of such a self homeomorphism of N_q. If t is a meridian for K then by Theorem 8 of Chapter 2 any other weight element is conjugate to tg or $(tg)^{-1}$ for some g in $\pi K''$. Now $\pi'' = \Gamma_q{}' = \langle z^q \rangle$ is contained in $\zeta\Gamma_q$ and so there is an automorphism of π carrying t to tg and which is the identity on π'. (Note however that t and tz are not conjugate). By Theorem 2 such an automorphism can be realized by a base point and orientation preserving self homeomorphism of $M(K)$, and so K is determined up to changes of orientation by π alone. (We may use the S^1-action on N_q to construct a base point, orientation and fibration preserving self homeomorphism of M realizing such an automorphism, instead of appealing to 4-dimensional surgery). If the characteristic polynomial is X^2-X+1 or X^2-3X+1 the monodromy is conjugate to its inverse (and the conjugating homeomorphism of N_q is base point and orientation preserving), so there is a self homeomorphism of M which reverses the orientation and the meridian, and such knots are +amphicheiral. On the other hand, the polynomials X^2+X-1 and X^2-X-1 are not symmetric and so the corresponding knots are not +amphicheiral. Since every automorphism of Γ_q is orientation preserving no such knot is −amphicheiral or invertible.

We shall show that for all but two of these knot groups the corresponding knots are determined by their exteriors. The exceptional cases correspond to the automorphisms with characteristic polynomial X^2-X+1 or X^2-3X+1. We shall treat these first.

Let T be the trefoil knot. Then $\pi\tau_6 T$ has a presentation $<t,u \mid tut = utu,\ t^6u = ut^6>$. A Reidemeister-Schreier rewriting process shows that π' is generated by $x = t^{-1}u$ and $y = tut^{-2}$ and is isomorphic to Γ. Moreover the meridianal automorphism is given by $\Theta = (A,(0,0))$ where $A = \begin{pmatrix} 1 & -1 \\ 1 & 0 \end{pmatrix}$. As an automorphism of *Nil*, Θ fixes the centre pointwise, and it has order 6. Therefore it induces a self homeomorphism θ of $N = N_1$ of order 6 which fixes pointwise a circle through the base point. The mapping torus $N\times_\theta S^1$ is homeomorphic to $M(\tau_6 T)$. (This can be seen without appealing to surgery). Note that $(\begin{pmatrix} 0 & 1 \\ 1 & 0 \end{pmatrix}, (0,0))$ is an involution of *Nil* which conjugates Θ to its inverse, and so M admits an orientation reversing involution. (In fact all the twist spins of a strongly invertible knot such as T are +amphicheiral [Li 1985]).

Let $\hat{M}$ be the (non-normal) covering space of M corresponding to the subgroup of π generated by t. Then $\hat{M}$ is the mapping torus $Nil\times_\Theta S^1$. We shall take $[0,0,0,0]$ as the base point of $\hat{M}$ and its image in M as the base point there. Any isotopy γ from $\gamma(0) = id_{Nil}$ to $\gamma(1) = \Theta$ (through homeomorphisms of *Nil*) determines a homeomorphism ρ_γ from $R^3\times S^1$ to $\hat{M}$ via $\rho_\gamma(r,s,t,e^{2\pi iu}) = [\gamma(u)([r,s,t]),u]$. We shall see that Θ is in fact isotopic to id_{Nil} through Lie group automorphisms.

Lemma 4 *There is a path γ in $Aut_{Lie}(Nil)$ from $\gamma(0) = id_{Nil}$ to $\gamma(1) = \Theta$ such that $\gamma(u)\Theta = \Theta\gamma(u)$ for all $0 \leqslant u \leqslant 1$.*

Proof The matrix $A(u) = uA+(1-u)I$ is invertible for all u in R, since the eigenvalues of A are complex. If we set $\gamma(u) = (A(u),\mu(u))$ then, on equating the second elements of the ordered pairs $\gamma(u)\Theta$ and $\Theta\gamma(u)$, we find that

$\mu(u)(A-I) = \mu(u)A-(detA)\mu(u)$ is uniquely determined. Since $det(A-I) = 1$ we can solve the equation uniquely for $\mu(u)$. Moreover, by the uniqueness, when $A(u) = I$ or A we must have $\mu(u) = (0,0)$. Thus γ is a path from id_{Nil} to Θ. □

Lemma 5 *Let $\Omega = (B,\nu)$ be an automorphism of Γ such that $\Omega\Theta = \Theta\Omega$. Then $\Omega = \Theta^r$ for some $r = 0, 1, 2, 3, 4$ or 5.*

Proof Since B is in $GL(2,Z)$, the equation $BA = AB$ readily implies that $B = A^r$ for some $r = 0, 1, 2, 3, 4$ or 5. As in Lemma 4 the second element ν is uniquely determined (it is in Z^2 since $A-I$ is in $GL(2,Z)$) and the uniqueness then implies that $\Omega = \Theta^r$. □

An immediate consequence of Lemma 5 is that if γ is any path as in Lemma 4 then it commutes with Ω. In [CS 1976] the meridional automorphisms were given by matrices A in $SL(3,Z)$, and the path $uA+(1-u)I$ remains in $GL(3,R)$ for all nonnegative u, provided that A has no negative eigenvalues. Such a path clearly commutes with any matrix B that commutes with A.

We now wish to provide an analogue for the key Proposition 2 of [CS 1976], which identifies the proper homotopy classes of lifts to $\hat{M}$ of certain self homotopy equivalences of M. Since M admits an involution that reverses the meridian it shall suffice to consider only homotopy equivalences which preserve the meridian.

Lemma 6 *Let h be a base point preserving self homotopy equivalence of M such that $h_*(t) = t$ in π, and let $\hat{h}$ be the unique lift of h to a base point preserving map from $\hat{M}$ to itself. Then $\hat{h}$ is properly homotopic to $id_{\hat{M}}$.*

Proof Let Ω be the automorphism of $\Gamma = \pi'$ induced by h_*. Then $\Omega\Theta = \Theta\Omega$, since $h_*(t) = t$. Considered as an automorphism of Nil, Ω determines self homeomorphisms ω of N and hence h_ω of M by $h_\omega([n,s]) = [\omega(n),s]$ for all $[n,s]$ in $M = N\times_\theta S^1$. (Note that h_ω is well defined since

$\Omega\Theta = \Theta\Omega$). Since $h_{\omega *} = h_*$ and M is aspherical, h and h_ω are homotopic. Therefore the lifts $\hat{h}$ and $\hat{h}_\omega$ to base point preserving maps of $\hat{M}$ are properly homotopic. Now $\hat{h}_\omega([l,s]) = [\Omega(l),s]$ for $[l,s]$ in $Nil\times_\Theta S^1$, and $\rho_\gamma^{-1}\hat{h}_\omega\rho_\gamma(l, e^{2\pi iu}) = (\gamma(u)^{-1}\Omega\gamma(u)(l), e^{2\pi iu}) = (\Omega(l), e^{2\pi iu})$, where γ is the path of Lemma 4. Thus $\rho_\gamma^{-1}\hat{h}_\omega\rho_\gamma = \Omega\times id_{S^1}$. Since Ω is orientation preserving, it is isotopic to id_{Nil}, and so $\hat{h}$ is properly homotopic to $id_{\hat{M}}$. □

From here on the argument is as in Section 3 of [CS 1976]. The next theorem is essentially their Proposition 3.

Theorem 4 *The 6-twist spin of the trefoil knot is not reflexive.*

Proof Let γ be the path from id_{Nil} to Θ constructed in Lemma 4 and ρ_γ the corresponding homeomorphism from $R^3\times S^1$ to $\hat{M}$. The composition with projection to M restricts to an embedding $r:D^3\times S^1 \to M$ onto a regular neighbourhood R of $\{0\}\times S^1$, which we may assume invariant under the orientation reversing involution of M described earlier. Let $\Sigma = (M-int\,R)\cup_r S^2\times D^2$ and $\Sigma_\tau = (M-int\,R)\cup_{r\tau} S^2\times D^2$ where τ is a self homeomorphism of $S^2\times S^1$ which preserves projection onto the S^1 factor and does not extend across $S^2\times D^2$ (cf. Chapter 1). (We shall also use the same symbol to denote the "radial" extension of τ to $R^3\times S^1$). If we fix an orientation for M then Σ and Σ_τ are naturally oriented also. Let K and $K_\tau = K^*$ be the knots obtained by restricting the natural inclusions j and j_τ of $S^2\times D^2$ into Σ and Σ_τ (respectively) to their core 2-spheres. Then up to change of orientations one of these is the 6-twist spin of the trefoil.

If there is a homeomorphism f from Σ to Σ_τ carrying K to K_τ (as unoriented submanifolds) then we may assume that f is orientation preserving, since K is +amphicheiral. It must then preserve the transverse and knot orientations also. By the uniqueness of regular neighbourhoods we may then arrange that $fj = j_\tau$, after an isotopy if necessary. Such a map f induces a self homeomorphism g of $M-int\,R$ such that $r^{-1}gr|S^2\times S^1 = \tau$ which extends radially across $D^3\times S^1$ to a self homeomorphism h of M. Clearly h fixes the meridian represented by $\{0\}\times S^1$ and so Lemma 6 applies. Thus the lift $\hat{h}$ is properly homotopic to $id_{\hat{M}}$. Since the self homeomorphisms τ and $\rho_\gamma^{-1}\hat{h}\rho_\gamma$ of $R^3\times S^1$ agree on $D^3\times S^1$, they are properly homotopic, and so τ is properly homotopic to the identity.

Now τ extends uniquely to a self homeomorphism τ of $S^3\times S^1$, and any such proper homotopy extends to a homotopy from τ to the identity. But this is known to be impossible. For let p be the projection of $S^3\times S^1$ onto S^3. The suspension of $p\tau$, restricted to the top cell of $S(S^3\times S^1) = S^2\vee S^4\vee S^5$ is the nontrivial element of $\pi_5(S^4)$, whereas the corresponding restriction of the suspension of p is trivial. (Cf. [CS 1976, Go 1976]). Therefore there is no such map f, and the knots are distinct. □

When $\pi' = \Gamma$ and the meridianal automorphism is of the form $\Theta = (D,(0,0))$ where D has characteristic polynomial X^2-3X+1 we can make a similar argument. Given an automorphism $\Omega = (B,\nu)$ commuting with Θ we find that unless Ω is the identity $detB$ is not an eigenvalue of B, and so we can find a path γ from id_{Nil} to Θ which commutes with Ω and which is linear in the $GL(2,R)$ entry. (However the path in general depends on Ω). The argument of Theorem 4 then applies to show that such knots are not determined by their exterior.

In the remaining cases with $\pi' = \Gamma_q$ we may assume that the meridianal automorphism has the form $\Theta = (D,(0,0))$ where $D = \begin{pmatrix}1 & 1\\ 1 & 0\end{pmatrix}$ or its inverse. In either case $\Omega = (-I,(-1,1))$ commutes with Θ and so determines a self homeomorphism h_ω of $M = N_q\times_\theta S^1$, which fixes pointwise the meridianal circle $\{0\}\times S^1$. The action of h_ω on the normal bundle may be detected by the induced action on $\hat{M}$. In each case we may construct an isotopy from Θ to $\Upsilon = (\begin{pmatrix}1 & 0\\ 0 & -1\end{pmatrix},(1,0))$ which commutes with Ω. Thus we may replace $\hat{M}$ by the mapping torus $Nil\times_\Upsilon S^1$. (Note also that under the canonical identification of Nil with R^3, the automorphisms Υ and Ω act linearly).

Let $\gamma(u) = \begin{pmatrix}1 & v(u)\\ 0 & R(u)\end{pmatrix}$, where $v(u) = (0,u)$ and $R(u)$ in $SL(2,R)$ is rotation through πu radians, for $0 \leqslant u \leqslant 1$. Then γ is a path from

$\gamma(0) = id_{Nil}$ to $\gamma(1) = \Upsilon$ which we may use to identify the mapping torus of Υ with $R^3 \times S^1$. In the "new coordinates" the action of h_ω is given by the map sending $(r,s,t,e^{2\pi iu})$ to $(\gamma(u)^{-1}\Omega\gamma(u)(r,s,t),e^{2\pi iu})$. It is not hard to see that the loop sending $e^{2\pi iu}$ in S^1 to $\gamma(u)^{-1}\Omega\gamma(u)$ in $SL(3,R)$ is freely homotopic to the loop $\gamma_1(u)^{-1}\Omega_1\gamma_1(u)$, where $\gamma_1(u) = \begin{pmatrix} 1 & 0 \\ 0 & R(u) \end{pmatrix}$ and $\Omega_1 = diag(-1,-1,1)$. Since the latter matrix product simplifies to $\begin{pmatrix} 1 & 0 \\ 0 & R(2u) \end{pmatrix}$, these loops are essential in $SL(3,R)$. Thus the self homeomorphism h_ω induces the twist τ on the normal bundle of the meridian, and so the knot is equivalent to its Gluck reconstruction.

The footnote in [CS 1976] may be fleshed out in a similar way. If K is a 2-knot with $\pi' = Z^3$ and its meridianal automorphism A has negative eigenvalues, then there is a matrix P in $GL(3,R)$ such that $PAP^{-1} = diag[\lambda_1,\lambda_2,\lambda_3]$, where λ_1 is positive and λ_2 and λ_3 are negative. If B commutes with A then PBP^{-1} commutes with PAP^{-1}, and so must be of the form $diag[\beta_1,\beta_2,\beta_3]$ (as the eigenvalues of A are distinct). On replacing B by $-B$ if necessary, we may assume that $detB = +1$. We may isotope PAP^{-1} linearly to $diag[1,-1,-1]$. If for every matrix B in $SL(3,Z)$ which commutes with A the minor $\beta_2\beta_3$ is positive then PBP^{-1} is isotopic to the identity through block diagonal matrices and we may argue as in Theorem 4 to conclude that K is not determined by its exterior. On the other hand, if there is a matrix B with $\beta_2\beta_3$ negative then we can argue as in the above two paragraphs to conclude that K *is* determined by its exterior. In Appendix B we shall relate this criterion to the question of whether all totally positive units in the ring of integers of the cubic number field determined by an eigenvalue of A are perfect squares.

In [HP 1988] it is shown that no fibred 2-knot with fibre having a geometric structure and monodromy of finite order greater than 2 can be reflexive. (The assumption that the fibre have a geometric structure is surely unnecessary).

The spherical case

When π' is finite, there is an infinite cyclic central subgroup of finite index in π, and the corresponding covering space of $M(K)$ is homeomorphic to $S^3 \times S^1$, by surgery over Z. In particular the universal cover $\widetilde{M}$ is homeomorphic to $S^3 \times R$, and if $\pi = Z$ the knot is trivial [Fr 1983]. The homotopy type of M is determined by π together with a k-invariant. However in general this does not determine M up to homeomorphism (cf. Theorem 6.3 of [Pl 1983]).

Determination of the range of simple homotopy types of such M seems difficult, largely because the *Nil* groups occuring in Waldhausen's sequence [Wa 1978] relating $Wh(\pi)$ to the K-theory of π' seem so intractable. We can however compute the Wall group $L_5^s(\pi)$ modulo 2-torsion and thus estimate the size of the structure set $S_{TOP}(M)$ fairly well.

For the Wall groups of π and π' are related by a Mayer-Vietoris sequence $L_5^s(\pi') \to L_5^s(\pi) \to L_4^u(\pi') \to L_4^s(\pi')$ [Ca 1973]. The right hand map is t_*-1, where t_* is induced by the meridianal automorphism of π', and the superscript u signifies that the torsion must lie in a certain subgroup of $Wh(\pi')$. Now $L_5^s(\pi')$ is a finite 2-group and $L_4^u(\pi') \sim L_4^s(\pi') \sim Z^R$ modulo 2-torsion, where R is the set of irreducible real representations of π' [W: Chapter 13A]. (Note that the latter correspond to the conjugacy classes of π' up to inversion. Cf. [Se: Section 12.4]). The meridianal automorphism t_* induces the identity on R, except when $\pi' = Q(1)$, in which case it permutes the 3 nontrivial 1-dimensional representations, and similarly when $\pi' = Q(1) \times Z/nZ$. With a little work it follows that $L_5^s(\pi)$ has rank $r+1$, $3(r+1)$, $3^{k-1}(5+7r)$ or $9(r+1)$ when $\pi' \cong P \times Z/(2r+1)Z$ with $P = 1$, $Q(1)$, $T(k)$ or I^* respectively.

As π is in Freedman's class, the whole of $L_5^s(\pi)$ acts on $S_{TOP}(M)$, and by Lemma 1 the action is transitive and the image of $\hat{I}_\pi$ acts trivially. This image is an infinite cyclic direct summand of $L_5^s(\pi)$. Since $[SM, G/TOP] = Z \oplus (Z/2Z)$, it follows from the above estimates that if $\pi' \neq 1$ then there are infinitely many s-cobordism classes of simple homotopy equivalences. Theorem 2 of Chapter 7 then implies that there are infinitely many distinct 4-manifolds simple homotopy equivalent to M.

When $\pi' = Z/(2r+1)Z$ we can be more explicit. In this case $L_5^s(\pi') = 0$ and the torsion condition in the Mayer-Vietoris sequence is $u = h$, so there is an isomorphism $L_5^s(\pi) \cong L_4^h(\pi')$. Moreover

$L_4^s(\pi') = Z^{r+1}$ and the natural map to $L_4^h(\pi')$ is an injection with cokernel a quotient of $H^2(Z/2Z;\tilde{K}_0(\pi'))$ [Ba 1978]. (This cokernel is trivial if $|\pi'|$ $(= 2r+1) \leqslant 13)$.

In the simplest nontrivial case, $\pi' = Z/3Z$ and π is the group of $\tau_2 T$, the 2-twist spin of the trefoil knot. If K is any knot with group π then $M(K)$ is homotopy equivalent to $M(\tau_2 T)$ (and the homotopy equivalence is simple if $Nil(Z[Z/3Z],-) = 0$).

Lemma 7 *Let K be a 2-knot. Then K has a Seifert hypersurface which contains no fake 3-cells.*

Proof By the standard obstruction theoretical argument and TOP transversality every 2-knot has a Seifert hypersurface. Thus K bounds a locally flat 3-submanifold V which has trivial normal bundle in S^4. If Δ is a homotopy 3-cell in V then $\Delta \times R = D^3 \times R$, by simply connected surgery, and the submanifold $\partial\Delta$ of $\partial(\Delta \times R) = \partial(D^3 \times R)$ is isotopic there to the boundary of a standard 3-cell in $D^3 \times R$ which we may use instead of Δ. □

The modification in Lemma 7 clearly preserves minimality, in the sense introduced in Chapter 7. (Every 2-knot has a Seifert hypersurface which is a once-punctured hyperbolic 3-manifold [Ru 1987], and so contains no fake 3-cells, but these are rarely minimal).

Theorem 5 *Let K be a 2-knot such that $\pi' = Z/3Z$, and which has a minimal Seifert hypersurface. Then K is fibred.*

Proof Let V be a minimal Seifert hypersurface for K. Then $\pi_1(V)$ is finite, and we may assume that V is irreducible, by Lemma 7. Let $\hat{V} = V \cup D^3$. As in Theorem 6 of Chapter 7, $\pi_1(\hat{V}) = \pi_1(W) = Z/3Z$, and $W = M - \hat{V} \times (-1,1)$ is an h-cobordism from $\hat{V}$ to itself. (We can avoid the appeal to Kaplansky's Lemma here by arguing as in Theorem 1 of Chapter 4 to show that the universal covering space of W has the homotopy type of S^3). Therefore $W = \hat{V} \times I$, for $S_{TOP}(\hat{V} \times I \text{ } rel \text{ } \partial)$ has just one element, by surgery over $Z/3Z$. (Note that $Wh(Z/3Z) = L_5(Z/3Z) = 0$).

Therefore M fibres over S^1, and so K is fibred also. □

If the 3-dimensional Poincaré conjecture were true then a closed 3-manifold with fundamental group $Z/3Z$ would be the lens space $L(3,1)$ [Ru 1986]. It would then follow that none of the infinitely many distinct 2-knots with $\pi' = Z/3Z$ other than $\tau_2 T$ could have a minimal Seifert surface. In particular there would be infinitely many closed 4-manifolds simple homotopy equivalent to $M(\tau_2 T)$ which do not fibre over S^1, and so we would have counter examples to a 4-dimensional analogue of Farrell's fibration theorem [Fa 1970]. (A similar argument is given in [Ru 1987]. Cf. also [We 1987] and [HT 1988]).

We do not know whether any of these knots (other than $\tau_2 T$) is PL in some PL structure on S^4. (An equivalent question is whether the corresponding manifold $M(K)$ has a PL structure).

It may be possible to construct other exotic 2-knots with π' finite by wrapping together the ends of nontrivial s-cobordisms, such as those recently found by Cappell and Shaneson [CS 1986].

Groups of finite geometric dimension 2

The group Φ may be viewed as an HNN extension $\Phi = Z*_\phi$, where ϕ maps Z onto its subgroup $2Z$. It is therefore in Waldhausen's class Cl, and so $Wh(\Phi) = 0$. This is also true of knot groups with π' free, and of all classical knot groups [Wa 1978]. (In fact there is no known example of a finitely presentable torsion free group which has nontrivial Whitehead group).

Theorem 6 *Let M be a closed orientable 4-manifold with fundamental group Φ. Then any homotopy equivalence $f:N \to M$ is homotopic to a homeomorphism.*

proof Although Φ is not square root closed accessible, Cappell's splitting theorem holds for it, by the remark on page 167 of [Ca 1976], so there is still an exact sequence $L_5(Z) \to L_5(Z) \to L_5(\Phi) \to L_4(Z) \to L_4(Z)$ where the extreme maps are essentially $1-L_*(\phi)$. (Cf. pages 498 *et seq.* of [Ca

1973], or [St 1987]). Since $L_5(\phi) = I_Z^{-1}\phi I_Z$, the left hand map is an isomorphism. The abelianization homomorphism from Φ to Z induces a map to the corresponding exact sequence for Z, considered as an HNN extension of the trivial group. A diagram chase now shows that the induced map from $L_5(\Phi)$ to $L_5(Z) = L_4(1) = Z$ is an isomorphism. Similarly $L_4(\Phi) = L_4(Z) = L_4(1)$. Now by Lemma 2 the map f is s-cobordant to id_M, and so f is homotopic to a homeomorphism. □

Corollary *A ribbon knot K with group Φ is determined up to orientation by the homotopy type of $M(K)$.*

Proof Since Φ is metabelian, there is an unique weight class up to inversion, so the knot exterior is determined by $M(K)$, and since K is a ribbon knot it is determined by its exterior. □

This corollary applies in particular to Examples 10 and 11 of Fox [Fo 1962]. Ribbon 2-knots are −amphicheiral, but no 2-knot with an asymmetric Alexander polynomial can be invertible. Thus as oriented knots these examples are distinct. Is there a 2-knot with group Φ which is not a ribbon knot?

Theorem 7 *Let K be a 2-knot such that π' is a free group of rank r. Then K is s-concordant to a fibred knot with closed fibre $\#^r(S^1\times S^2)$, which is determined (among such fibred knots) up to changes of orientations by π together with the weight orbit of a meridian. Moreover any such fibred knot is reflexive and homotopy ribbon.*

Proof By Theorem 5 of Chapter 7 $M(K)$ is simple homotopy equivalent to a PL 4-manifold N which fibres over S^1 with fibre $\#^r(S^1\times S^2)$, and which is determined among such manifolds by its group. As the group π is square root closed accessible, abelianization induces an isomorphism of $L_*(\pi)$ with $L_*(Z)$. Therefore by Lemma 2 there is an s-cobordism Z from M to N. We may embed an annulus $A = S^1\times[0,1]$ in Z so that $M\cap A = S^1\times\{0\}$ is a meridian for K and $N\cap A = S^1\times\{1\}$. Surgery on A in Z then gives an s-concordance from K to such a fibred knot K_1, which is reflexive [Gl

1962] and homotopy ribbon [Co 1983]. □

Corollary *The 0-spin of an invertible fibred knot is determined up to s-concordance by its group together with the weight orbit of a meridian.*

Proof The 0-spin of a classical knot is -amphicheiral and reflexive. □

If the 3-dimensional Poincaré conjecture holds then every fibred 2-knot with π' free is homotopy ribbon [Co 1983]. Is every such group the group of a ribbon knot?

Appendix A Four-Dimensional Geometries and Smooth 2-Knots

A *smooth* 2-knot is a smooth embedding K of S^2 into a smooth homotopy 4-sphere Σ. The manifold $M(K)$ obtained by surgery on K is then a smooth orientable 4-manifold, and if it fibres smoothly over S^1 we shall say that K is a smoothly fibred 2-knot. In a number of cases M supports a 4-dimensional geometry. There are in fact 19 classes of 4-dimensional geometries. In the light of the role that algebraic surfaces are currently playing in 4-dimensional differential topology it is of interest that some of these manifolds in fact admit nonsingular complex analytic structures. (These cannot be algebraic or even Kähler, as $\beta_1(M)$ is odd. See [Wa 1986] and the references there, in particular to the work of Filipkiewicz, Inoue, Kato and Kodaira, for more details on 4-dimensional geometries and complex surfaces).

If K is a branched twist spin of a torus knot or a simple knot then M fibres over S^1 with fibre a closed irreducible 3-manifold with a geometric structure (of type S^3, $\widetilde{SL}$, Nil^3, H^3 or E^3) and monodromy having finite order and nonempty fixed point set. The manifold M then has a finite cover which has the corresponding 4-dimensional product geometry ($S^3 \times E^1$, etc.), and so M admits this geometry alone, if any at all. We shall show that the only other geometries that may be realized by such 4-manifolds $M(K)$ are Sol_0^4, Sol_1^4 and $Sol_{m,m\pm 1}^4$ for $2m\pm 1 \geqslant 11$, and possibly H^4 or $H^2(C)$ (although we know of no examples of the latter types).

A closed 4-manifold admits at most one geometry, and homotopy equivalent geometrizable manifolds must have the same geometry, by Theorem 10.1 of [Wa 1986]. The 4-dimensional geometries not already mentioned are $S^2 \times S^2$, $S^2 \times E^2$, $S^2 \times H^2$, $E^2 \times H^2$, $H^2 \times H^2$, S^4, $P^2(C)$, Nil^4 and F^4. The last of these cannot be realized by any closed 4-manifold, and so cannot occur here. Since the universal cover $\widetilde{M}$ is an open 4-manifold the geometries $S^2 \times S^2$, S^4 and $P^2(C)$ cannot occur. If $\widetilde{M}$ is not contractible then the geometry must be a product geometry, $S^2 \times E^2$, $S^2 \times H^2$ or $S^3 \times E^1$. Of these, only $S^3 \times E^1$ admits discrete uniform actions of groups with infinite cyclic abelianization.

All of the remaining geometries have contractible models. Most of the geometries of solvable type can be realized by smooth 2-knots with

M aspherical and π solvable. The results below in such cases follow from our determination of the virtually torsion free solvable 2-knot groups in conjunction with the description of the discrete uniform subgroups of the isometry groups of the solvable 4-dimensional geometries given in Section 2 of [Wa 1986].

If M has a geometry of type $S^3 \times E^1$ then π' is finite and $\widetilde{M}$ is diffeomorphic to $S^3 \times R = R^4 - \{0\}$. If M is a complex surface (compact complex analytic manifold of complex dimension 2) and π' is finite then M is a Hopf surface, i.e. $\widetilde{M}$ is analytically isomorphic to $C^2 - \{0\}$ (Kodaira). Moreover M is then determined up to diffeomorphism by π, and π' must be cyclic, $T(1)$ or I^* (Kato).

If M has a geometry of type E^4 (i.e. is flat) then it is determined up to diffeomorphism by π, which is virtually Z^4. Since π must be isomorphic to $G(+)$ or $G(-)$, there are just two such manifolds. Neither supports a complex structure.

If M has a geometry of type $Nil^3 \times E^1$ then it is fibred with fibre of Nil^3 type and monodromy of finite order, and so is diffeomorphic to $M(K_1)$ where K_1 is either the 2-twist spin of a Montesinos knot $K(0 \mid b;(3,1),(3,1),(3,\pm 1))$ for some odd b or the 6-twist spin of the trefoil knot. These manifolds admit complex surface structures as secondary Kodaira surfaces.

The discrete uniform subgroups of $Sol^4_{m,n}$ have infinite cyclic abelianization if and only if $m-n = \pm 1$. The manifold $M(K)$ has a geometry of type $Sol^4_{m,m\pm 1}$ if and only if K is a Cappell-Shaneson knot and all the roots of its Alexander polynomial are positive, in which case $2m \pm 1 \geqslant 11$. If K is a Cappell-Shaneson 2-knot whose Alexander polynomial has one positive root and two negative roots then $M(K)$ admits no geometric structure, but its 2-fold cover has a geometry of type $Sol^4_{m,n}$ for some m, n. These manifolds do not support complex structures.

However the six distinct manifolds arising from the Cappell-Shaneson knots whose Alexander polynomial has only one real root have geometry of type Sol^1_0, and all admit complex structures as Inoue surfaces of type S_M. (Cf. Section 9 of [Wa 1986]). These manifolds may be distinguished by the Alexander polynomials of the corresponding 2-knots (cf. Table 1 of [AR 1984]).

If M has a geometry of type Sol_1^4 then it is fibred, with fibre a coset space of Nil^3 and monodromy of infinite order. One such manifold has a family of complex structures as an Inoue surface of type $S_{N,t}^+$, depending on a complex parameter t; the others form an infinite family, all having complex structures as Inoue surfaces of type S_N^-. This family has a "universal" member: each surface is a quotient of the universal surface by a free action of a finite cyclic group of odd order.

Since any discrete uniform subgroup of Nil^4 has abelianization of rank at least 2, this geometry cannot occur.

If $\widetilde{M}$ is contractible but π is not solvable then the geometry must be one of H^4, $H^2(C)$, $H^3{\times}E^1$, $\widetilde{SL}{\times}E^1$, $H^2{\times}H^2$ or $E^2{\times}H^2$. The last three of these do not admit discrete uniform actions of any group with infinite cyclic abelianization, and so cannot occur. No manifold with geometry of type H^4 or $H^3{\times}E^1$ can be a complex surface. (According to Bogomolov, the Hopf and Inoue surfaces are the only nonelliptic complex surfaces with $\beta_1 = 1$ and $\beta_2 = 0$).

Any $M(K)$ admitting one of the above geometries other than H^4 or $H^2(C)$ is fibred, with fibre a geometric 3-manifold. The TOP classification (up to Gluck reconstruction) of the knots of type $Nil^3{\times}E^1$, E^4, Sol_o^4, Sol_1^4 and $Sol_{m,m\pm1}^4$ given in Chapter 8 then gives also the smooth classification. In the present context however it is natural to require that the weight class be represented by a cross-section of the (geometric) fibration. (Recall also that by Plotnick's theorem any fibred 2-knot whose closed monodromy has finite order and nonempty fixed point set is a branched twist spin, if the 3-dimensional Poincaré conjecture is true).

Appendix B Reflexive Cappell–Shaneson 2–Knots

Let A be a matrix in $SL(3,Z)$ such that $det(A-I) = \pm 1$. On replacing A by its inverse if necessary, we may assume that $det(A-I) = 1$. The characteristic polynomial of A is then $f_a(X) = X^3 - aX^2 + (a-1)X - 1$, where a is the trace of A. It is easy to see that f_a is irreducible and has either 0 or 2 (distinct) negative roots. We shall assume that f_a has one positive root λ_1 and two negative roots λ_2 and λ_3. (This is so if and only if a is negative). Since the eigenvalues of A are distinct and real, there is a matrix P in $GL(3,R)$ such that $\tilde{A} = PAP^{-1}$ is the diagonal matrix $diag[\lambda_1,\lambda_2,\lambda_3]$. If B in $GL(3,Z)$ commutes with A then $\tilde{B} = PBP^{-1}$ commutes with $\tilde{A}$ and so must also be diagonal (as the λ_i's are distinct). Suppose that $\tilde{B} = diag[\beta_1,\beta_2,\beta_3]$. Then the criterion of the footnote in [CS 1976] for a 2-knot with $\pi' = Z^3$ and meridional automorphism A to be determined by its exterior is that $\beta_2\beta_3$ should be negative. On replacing B by $-B$ if necessary, we may assume that $detB = 1$ and the criterion then becomes $\beta_1 < 0$.

Let F be the field $Q[X]/(f_a)$ and let O_F be the ring of integers in F. We may view Q^3 as a $Q[X]$-module, and hence as a 1-dimensional F-vector space via the action of A. If B in $GL(3,Z)$ commutes with A then it induces an automorphism of this vector space which preserves a lattice, and so determines a unit $u(B)$ of O_F. Moreover, $detB = N_{F/Q}(u(B))$. Every unit which maps the image of $Z[X]/(f_a)$ to itself arises in this way. In particular, this is so if $O_F = Z[X]/(f_a)$. (Note however that, for instance, $f_{-22} = (X-2)^3+7(X-2)(4X+3)+7^2$ is in the square of a maximal ideal of $Z[X]$ and so the ring $Z[X]/(f_{-22})$ is not integrally closed [Hi 1984]).

Let σ be the embedding of F in R which sends the image of X in F to λ_1. Then if P and B are as above, we must have $\sigma(u(B)) = \beta_1$. Thus if the criterion of [CS 1976] holds, there is a unit u in O_F such that $N_{F/Q}(u) = 1$ and $\sigma(u) < 0$. Conversely, if there is such a unit, and if $O_F = Z[X]/(f_a)$ then there is such a matrix B.

We may reduce the question of the existence of such a unit to a standard problem of number theory as follows. Let U be the group of all units of O_F, let U^σ be the subgroup of units whose image under σ is positive, let U^+ be the subgroup of totally positive units, and let U^2 be the subgroup of squares of units. Then $U \cong (\pm 1)\times Z^2$, since F is a totally real cubic number field, and so U/U^2 has order 8. The unit -1 has norm -1 and the element λ_1 is a unit of norm $+1$ in U^σ which is not totally positive (since its conjugates are λ_2 and λ_3). It is now easy to check that there is a unit of norm $+1$ which is not in U^σ if and only if $U^+ = U^2$, i.e. if and only if every totally positive unit is a perfect square in U.

When the discriminant of f_a is a perfect square in Z, the field F is Galois over Q, with group $Z/3Z$. If moreover the class number of F is odd, then $U^+ = U^2$, by [AF 1967]. In particular, if A has trace -1 then the characteristic polynomial has discriminant 49, the ring $Z[X]/(f_{-1})$ is the full ring of integers, and the class number is 1 (cf. [AR 1984]), so the corresponding 2-knot is determined by its group, up to a reflection. (In this case the polynomial f_{-1} determines the conjugacy class of A, by the theorem of Latimer and MacDuffee [New: page 52], and so determines the group among metabelian 2-knot groups).

Some Open Questions

In the tradition of [N] and [GK], we list here a number of questions which we have not been able to settle. Of course some of these questions are well known to be very difficult.

1. Is the Disk Embedding Theorem valid over arbitrary fundamental groups? In particular, are s-concordant 2-knots isotopic?

2. What can be said about *smooth* 2-knots? In particular, is every fibred 2-knot isotopic to one which is smooth in the standard smoothing of S^4?

3. When is a 2-knot fibred? Is this so if it has a minimal Seifert surface and the knot group has finitely generated commutator subgroup?

4. Is every PD_3-group with nontrivial centre the fundamental group of a Seifert fibred 3-manifold?

5. Is every PD_3-complex X such that $e(\pi_1(X)) = \infty$ a connected sum?

6. Let N be a 3-manifold, and Δ a homotopy 3-cell in N such that $N-\Delta$ contains no fake 3-cells. Can every self homeomorphism of N be isotoped so as to leave Δ invariant? In particular, is this so if N is either aspherical or has free fundamental group?

7. Let M be a closed 4-manifold with $\chi(M) = 0$ and $\pi_2(M) = 0$. Must $\pi_3(M)$ be finitely generated? (If so then $e(\pi_1(M)) = 1$ or 2, and so either M is aspherical or it is finitely covered by $S^3 \times S^1$).

8. Let M be a closed 4-manifold with $\chi(M) = 0$ and such that $\pi_1(M)$ has a finitely presentable normal subgroup H with quotient Z. Is the corresponding covering space a PD_3-complex? In particular, if $e(H) = 1$ must M be aspherical?

9. Let M be a closed orientable 4-manifold and $f:M \to S^1$ a map which induces an epimorphism on fundamental groups. When can f be homotoped

to a map transverse over 1 in S^1 and such that one (or both) of the pushoff maps from $F = f^{-1}(1)$ into $M - F\times(-\varepsilon,\varepsilon)$ induces a monomorphism on fundamental groups?

10. Is the homotopy type of $M(K)$ uniquely determined by πK if $c.d.\pi = 2$? In particular, is this so when $\pi \cong \Phi$?

11. Is there a 2-knot K such that $\tilde{M}(K)$ is contractible but *not* homeomorphic to R^4?

12. Is every 2-knot uniquely factorizable as a sum of irreducible knots?

13. Is every homotopy ribbon 2-knot a ribbon knot? In particular, is every 2-knot group π with π' free the group of a (fibred) ribbon knot?

14. If the group π of a fibred 2-knot has deficiency 1, must π' be free?

15. Show that if $r > 2$ the r-twist spin of a nontrivial 1-knot is never reflexive.

16. If the centre of a 2-knot group is nontrivial, must it be $Z/2Z$, Z, $Z\oplus(Z/2Z)$ or Z^2? Is the centre of the group of a 2-link with more than one component always trivial?

17. Must a virtually locally-finite by solvable 2-knot group with one end be torsion free?

18. Is there a 2-knot group which has a rank 2 abelian normal subgroup contained in its commutator subgroup?

19. If a 2-knot group π has a rank 1 abelian normal subgroup A such that $e(\pi/A) = 1$ must it be a PD_4^+-group?

20. Is every 3-knot group which is a PD_4^+-group in which some nonzero power of a weight element is central the group of a branched twist spin of a prime classical knot?

21. Does either of the conditions "$c.d.G = 2$" and "$def\ G = \mu$" imply the other, when G is a finitely presentable group with $G/G' \cong Z^{\mu}$ and $H_2(G;Z) = 0$?

22. If the commutator subgroup of a 2-knot group is finitely generated, must it be finitely presentable? coherent?

23. Which Λ-modules and torsion pairings are realized by $H_1(X(K);\Lambda)$ for some 2-knot K?

24. Find criteria for a 2-knot to be doubly slice. In particular, must a 2-knot with π'/π'' finite and hyperbolic torsion pairing be TOP doubly slice? A related question is: does every rational homology 3-sphere with hyperbolic linking pairing embed (TOP locally flat) in S^4?

25. What can be said about the fibred 2-knots which are the links of isolated singularities of polynomial maps from R^5 to R^2?

26. Determine the smooth 2-knots K such that $M(K)$ admits a 4-dimensional geometric structure.

27. Let G be an extension of Z by the fundamental group of a closed orientable 3-manifold which is hyperbolic or Haken. Is $Wh(G) = 0$?

28. Compute $Wh(\pi)$ and $L_5^s(\pi)$ when π is an extension of Z by a finite normal subgroup with cohomological period 4.

29. If G is a finite group with cohomological period 4 there is a (finite) orientable PD_4-complex with fundamental group $G\times Z$ and Euler characteristic 0. When is there such a closed 4-manifold? (Such a manifold can fibre over S^1 only if G is a 3-manifold group).

30. Determine the finite homology 4-sphere groups. In particular, is $I^*\times I^*$ one such?

References

We shall abbreviate "Lecture Notes in Mathematics (-), Springer-Verlag, Berlin-Heidelberg-New York" as "LN (-)".

[AR 1984] Aitchison, I.R. and Rubinstein, J.H. Fibered knots and involutions on homotopy spheres, in [GK], 1-74.

[AS 1988] Aitchison, I.R. and Silver, D.S. On certain fibred ribbon disc pairs, Trans. Amer. Math. Soc. 306 (1988), 529-551.

[AF 1967] Armitage, J.V. and Fröhlich, A. Class numbers and unit signatures, Mathematika 14 (1967), 94-98.

[Ar 1925] Artin, E. Zur Isotopie Zwei-dimensionaler Flächen im R_4, Abh. Math. Sem. Univ. Hamburg 4 (1925), 174-177.

[AY 1981] Asano, K. and Yoshikawa, K. On polynomial invariants of fibred 2-knots, Pacific J. Math. 97 (1981), 267-269.

[Ba 1978] Bak, A. The computation of even dimensional surgery groups of odd torsion groups, Comm. Alg. 6 (1978), 1393-1458.

[Ba 1980] Barge, J. An algebraic proof of a theorem of J.Milnor, in *Topology Symposium Siegen* 1979, (edited by U.Koschorke and W.D.Neumann), LN 788 (1980), 396-398.

[Ba 1980'] Barge, J. Dualité dans les revêtements galoisiens, Inventiones Math. 58 (1980), 101-106.

[Ba 1986] Baues, H.J. The homotopy type of 4-dimensional CW-complexes, preprint, Max-Planck-Institut für Mathematik, Bonn (1986).

[BB 1976] Baumslag, G. and Bieri, R. Constructable solvable groups, Math. Z. 151 (1976), 249-257.

[BHK 1981] Bayer, E., Hillman, J.A. and Kearton, C. The factorization of simple knots, Math. Proc. Cambridge Philos. Soc. 90 (1981), 495-506.

[Be 1984] Bedient, R.E. Double branched covers and pretzel knots, Pacific J. Math. 112 (1984), 265-272.

[B] Bieri, R. *Homological Dimension of Discrete Groups*, Queen Mary College Mathematical Notes, London (1976).

[BS 1978] Bieri, R. and Strebel, R. Almost finitely presented soluble groups, Commentarii Math. Helvetici 53 (1978), 258-278.

[BS 1979] Bieri, R. and Strebel, R. Soluble groups with coherent group rings, in [W'], 235-240.

[BO 1986] Boileau, C.M. and Otal, J-P. Groupe des difféotopies de certaines variétés de Seifert,

C.R. Acad. Sci. Paris 303 (1986), 19–22.

[BG 1984] Brooks, R. and Goldman, W. Volumes in Seifert space, Duke Math. J. 51 (1984), 529–545.

[Br 1967] Browder, W. Diffeomorphisms of 1-connected manifolds, Trans. Amer. Math. Soc. 128 (1967), 155–163.

[Br 1972] Browder, W. Poincaré spaces, their normal fibrations and surgery, Inventiones Math. 17 (1972), 191–202.

[Br 1975] Brown, K.S. Homological criteria for finiteness, Commentarii Math. Helvetici 50 (1975), 129–135.

[BG 1985] Brown, K.S. and Geoghegan, R. Cohomology with free coefficients of the fundamental group of a graph of groups, Commentarii Math. Helvetici 60 (1985), 31–45.

[BM 1970] Burde, G. and Murasugi, K. Links and Seifert fibre spaces, Duke Math. J. 37 (1970), 89–93.

[BZ 1966] Burde, G. and Zieschang, H. Eine Kennzeichnung der Torusknoten, Math. Ann. 167 (1966), 169–176.

[BZ] Burde, G. and Zieschang, H. *Knots*, Studies in Mathematics 5, W. de Gruyter, Berlin–New York (1985).

[CR 1980] Campbell, C.M. and Robertson, E.F. A deficiency zero presentation for $SL(2,p)$, Bull. London Math. Soc. 12 (1980), 17–20.

[Ca 1970] Cappell, S.E. Superspinning and knot complements, in *Topology of Manifolds* (edited by J.C.Cantrell and C.H.Edwards, Jr), Markham Publishing Co., Chicago (1970), 358–383.

[Ca 1973] Cappell, S.E. Mayer–Vietoris sequences in Hermitean K-theory, in *Hermitean K-Theory and Geometric Applications* (edited by H.Bass), LN 343 (1975), 478–512.

[Ca 1976] Cappell, S.E. A splitting theorem for manifolds, Inventiones Math. 33 (1976), 69–170.

[CS 1976] Cappell, S.E. and Shaneson, J.L. There exist inequivalent knots with the same complement, Ann. Math. 103 (1976), 349–353.

[CS 1976'] Cappell, S.E. and Shaneson, J.L. Some new four-manifolds, Ann. Math. 104 (1976), 61–72.

[CS 1986] Cappell, S.E. and Shaneson, J.L. On 4-dimensional s-cobordisms, J. Diff. Geometry 22 (1985), 97–115.

[CE] Cartan, H. and Eilenberg, S. *Homological Algebra*,

Princeton University Press, Princeton (1953).

[Co 1983] Cochran, T. Ribbon knots in S^4, J. London Math. Soc. 28 (1983), 563-576.

[C] Cohn, P.M. *Free Rings and their Relations*, Academic Press, New York-London (1971).

[CR 1977] Conner, P.E. and Raymond, F. Deforming homotopy equivalences to homeomorphisms in aspherical manifolds, Bull. Amer. Math. Soc. 83 (1977), 36-85.

[Da 1983] Davis, M.W. Groups generated by reflections and aspherical manifolds not covered by Euclidean space, Ann. Math. 117 (1983), 293-324.

[Du 1983] Dunbar, W. Geometric orbifolds, preprint, Rice University (1983).

[Du 1985] Dunwoody, M.J. The accessibility of finitely presented groups, Inventiones Math. 81 (1985), 449-457.

[DF 1987] Dunwoody, M.J. and Fenn, R.A. On the finiteness of higher knot sums, Topology 26 (1987), 337-343.

[DV 1973] Dyer, E. and Vasquez, A.T. The sphericity of higher dimensional knots, Canadian J. Math. 25 (1973), 1132-1136.

[Dy 1987] Dyer, M.N. Localization of group rings and applications to 2-complexes, Commentarii Math. Helvetici 62 (1987), 1-17.

[Dy 1987'] Dyer, M.N. Euler characteristics of groups, Quarterly J. Math. Oxford 38 (1987), 35-44.

[Ec 1976] Eckmann, B. Aspherical manifolds and higher-dimensional knots, Commentarii Math. Helvetici 51 (1976), 93-98.

[Ec 1986] Eckmann, B. Cyclic homology of groups and the Bass conjecture, Commentarii Math. Helvetici 61 (1986), 193-202.

[EM 1980] Eckmann, B. and Müller, H. Poincaré duality groups of dimension two, Commentarii Math. Helvetici 55 (1980), 510-520.

[EM 1982] Eckmann, B. and Müller, H. Plane motion groups and virtual Poincaré duality groups of dimension two, Inventiones Math. 69 (1982), 293-310.

[Fa 1977] Farber, M.Sh. Duality in an infinite cyclic covering and even-dimensional knots, Math. USSR Izvestija 11 (1977), 749-781.

[Fa 1983] Farber, M.Sh. The classification of simple knots, Russian Math. Surveys 38:5 (1983), 63-117.

[Fa 1970] Farrell, F.T. The obstruction to fibering a manifold over a circle, in *Actes, Congres internationale des Mathematiciens*, Nice (1970), tome 2, 69–72.

[FH 1981] Farrell, F.T. and Hsiang, W.C. The Whitehead group of poly-(finite or cyclic) groups, J. London Math. Soc. 24 (1981), 308–324.

[FH 1983] Farrell, F.T. and Hsiang, W.C. Topological characterization of flat and almost flat riemannian manifolds M^n ($n \neq 3,4$), Amer. J. Math. 105 (1983), 641–672.

[FJ 1988] Farrell, F.T and Jones, L.E. Topological rigidity for hyperbolic manifolds, Bull. Amer. Math. Soc. 19 (1988), 277–282.

[FKV 1987] Finashin, S.M., Kreck, M. and Viro, O.Ya. Exotic knottings of surfaces in the 4-sphere, Bull. Amer. Math. Soc. 17 (1987), 287–290.

[F] Fort, M.K., Jr. (editor) *Topology of 3-Manifolds and Related Topics*, Prentice-Hall, Englewood Cliffs, N.J. (1962).

[Fo 1962] Fox, R.H. A quick trip through knot theory, in [F], 120–167.

[Fo 1966] Fox, R.H. Rolling, Bull. Amer. Math. Soc. 72 (1966), 162–164.

[Fr 1975] Freedman, M.H. Automorphisms of circle bundles over surfaces, in *Geometric Topology* (edited by L.C.Glaser and T.B.Rushing), LN 438 (1975), 212–214.

[Fr 1982] Freedman, M.H. The topology of four-dimensional manifolds, J. Diff. Geometry 17 (1982), 357–453.

[Fr 1983] Freedman, M.H. The disk theorem for four-dimensional manifolds, in *Proceedings of the International Congress of Mathematicians, Warsaw* (1983), vol. 1, 647–663.

[FQ] Freedman, M.H. and Quinn, F. *Topology of 4-Manifolds*, Princeton University Press, Princeton (in preparation).

[Ga 1987] Gabai, D. Foliations and the topology of 3-manifolds. III, J. Diff. Geometry 26 (1987), 479–536.

[GM 1986] Geoghegan, R. and Mihalik, M.L. A note on the vanishing of $H^n(G;Z[G])$, J. Pure Appl. Alg. 39 (1986), 301–304.

[GR 1962] Gerstenhaber, M. and Rothaus, O. The solution of sets of equations in groups, Proc. Nat. Acad. Sci. USA 48 (1962), 1531–1533.

[Gi 1979] Gildenhuys, D. Classification of soluble groups of cohomological dimension two, Math. Z. 166 (1979), 21–25.

[GS 1981] Gildenhuys, D. and Strebel, R. On the cohomological dimension of soluble groups, Canadian Math. Bull. 24 (1981), 385-392.

[Gl 1962] Gluck, H. The embedding of two-spheres in the four-sphere, Trans. Amer. Math. Soc. 104 (1962), 308-333.

[GK 1978] Goldsmith, D.L. and Kauffmann, L.H. Twist spinning revisited, Trans. Amer. Math. Soc. 239 (1978), 229-251.

[Go 1988] Gompf, R.E. On Cappell-Shaneson 4-spheres, preprint, University of Texas at Austin (1988).

[GM 1978] Gonzalez-Acuña, F. and Montesinos, J.M. Ends of knot groups, Ann. Math. 108 (1978), 91-96.

[Go 1976] Gordon, C.McA. Knots in the 4-sphere, Commentarii Math. Helvetici 51 (1976), 585-596.

[Go 1981] Gordon, C.McA. Ribbon concordance of knots in the 3-sphere, Math. Ann. 257 (1981), 157-170.

[GK] Gordon, C.McA. and Kirby, R.C. (editors) *Four-Manifold Theory*, CONM 35, American Mathematical Society, Providence (1984).

[GL 1984] Gordon, C.McA. and Litherland, R.A. Incompressible surfaces in branched coverings, in *The Smith Conjecture* (edited by J.Morgan and H.Bass), Academic Press, New York-London (1984), 139-152.

[GL 1988] Gordon, C.McA. and Luecke, J. Knots are determined by their complements, preprint, University of Texas at Austin (1988).

[Go 1965] Gottlieb, D.H. A certain subgroup of the fundamental group, Amer. J. Math. 87 (1965), 840-856.

[Go 1979] Gottlieb, D.H. Poincaré duality and fibrations, Proc. Amer. Math. Soc. 76 (1979), 148-150.

[Gu 1971] Gutiérrez, M.A. Homology of knot groups: I. Groups with deficiency one, Bol. Soc. Mat. Mexico 16 (1971), 58-63.

[Gu 1972] Gutiérrez, M.A. Boundary links and an unlinking theorem, Trans. Amer. Math. Soc. 171 (1972), 491-499.

[Gu 1978] Gutiérrez, M.A. On the Seifert surface of a 2-knot, Trans. Amer. Math. Soc. 240 (1978), 287-294.

[Gu 1979] Gutiérrez, M.A. Homology of knot groups: III. Knots in S^4, Proc. London Math. Soc. 39 (1979), 469-487.

[HK 1988] Hambleton, I. and Kreck, M. On the classification of topological 4-manifolds with finite fundamental group, Math. Ann. 280 (1988), 85-104.

[HK 1978] Hausmann, J-C. and Kervaire, M. Sous-groupes dérivés des groupes de noeuds, L'Enseignement Math. 24 (1978), 111-123.

[HK 1978'] Hausmann, J-C. and Kervaire, M. Sur le centre des groupes de noeuds multidimensionels, C.R. Acad. Sci. Paris 287 (1978), 699-702.

[HW 1985] Hausmann, J-C. and Weinberger, S. Caractéristiques d'Euler et groupes fondamentaux des variétés de dimension 4, Commentarii Math. Helvetici 60 (1985), 139-144.

[He 1977] Hendriks, H. Obstruction theory in 3-dimensional topology: an extension theorem, J. London Math. Soc. 16 (1977), 160-164. Corrigendum *ibid.* 18 (1978), 192.

[H] Hillman, J.A. *Alexander Ideals of Links*, LN 895 (1981).

[Hi 1977] Hillman, J.A. High dimensional knot groups which are not two-knot groups, Bull. Austral. Math. Soc. 16 (1977), 449-462.

[Hi 1980] Hillman, J.A. Orientability, asphericity and two-knots, Houston J. Math. 6 (1980), 67-76.

[Hi 1981] Hillman, J.A. Aspherical four-manifolds and the centres of two-knot groups, Commentarii Math. Helvetici 56 (1981), 465-473. Corrigendum, *ibid.* 58 (1983), 166.

[Hi 1984] Hillman, J.A. Polynomials determining Dedekind domains, Bull. Austral. Math. Soc. 29 (1984), 167-175.

[Hi 1985] Hillman, J.A. Seifert fibre spaces and Poincaré duality groups, Math. Z. 190 (1985), 365-369.

[Hi 1986] Hillman, J.A. Abelian normal subgroups of two-knot groups, Commentarii Math. Helvetici 61 (1986), 122-148.

[Hi 1987] Hillman, J.A. The kernel of integral cup product, J. Austral. Math. Soc. 43 (1987), 10-15.

[Hi 1988] Hillman, J.A. Two-knot groups with torsion free abelian normal subgroups of rank two, Commentarii Math. Helvetici 63 (1988), 664-671.

[Hi 198?] Hillman, J.A. A homotopy fibration theorem in dimension four, Topology Appl., to appear.

[Hi 198?'] Hillman, J.A. The algebraic characterization of the exteriors of certain 2-knots, Inventiones Math., to appear.

[HP 1988] Hillman, J.A. and Plotnick, S.P. Geometrically fibred two-knots, preprint, Macquarie University and the State University of New York at Albany (1988).

[Hi 1979] Hitt, L.R. Examples of higher-dimensional slice knots which are not ribbon knots, Proc. Amer. Math. Soc. 77 (1979), 291-297.

[Ho 1942] Hopf, H. Fundamentalgruppe und zweite Bettische Gruppe, Commentarii Math. Helvetici 14 (1941/2), 257-309.

[Ho 1982] Howie, J. On locally indicable groups, Math. Z. 180 (1982), 445-461.

[Ho 1985] Howie, J. On the asphericity of ribbon disc complements, Trans. Amer. Math. Soc. 289 (1985), 281-302.

[Ka 1980] Kanenobu, T. 2-knot groups with elements of finite order, Math. Sem. Notes Kobe University 8 (1980), 557-560.

[Ka 1983] Kanenobu, T. Groups of higher dimensional satellite knots, J. Pure Appl. Alg. 28 (1983), 179-188.

[Ka 1983'] Kanenobu, T. Fox's 2-spheres are twist spun knots, Mem. Fac. Sci. Kyushu University (Ser. A) 37 (1983), 81-86.

[Ka 1988] Kanenobu, T. Deforming twist spun 2-bridge knots of genus one, Proc. Japan Acad. (Ser. A) 64 (1988), 98-101.

[K] Kaplansky, I. *Fields and Rings*, Chicago Lectures in Mathematics, Chicago University Press, Chicago-London (1969).

[Ka 1969] Kato, M. A concordance classification of PL homeomorphisms of $S^p \times S^q$, Topology 8 (1969), 371-383.

[Ke 1975] Kearton, C. Blanchfield duality and simple knots, Trans. Amer. Math. Soc. 202 (1975), 141-160.

[Ke 1965] Kervaire, M.A. Les noeuds de dimensions supérieures, Bull. Soc. Math. France 93 (1965), 225-271.

[Ke 1965'] Kervaire, M.A. On higher dimensional knots, in *Differential and Combinatorial Topology* (*A Symposium in Honor of Marston Morse*) (edited by S.S.Cairns), Princeton University Press, Princeton (1965), 105-109.

[Ke 1969] Kervaire, M.A. Smooth homology spheres and their fundamental groups, Trans. Amer. Math. Soc. 144 (1969), 67-72.

[KS 1975] Kirby, R.C. and Siebenmann, L.C. Normal bundles for codimension 2 locally flat imbeddings, in *Geometric Topology* (edited by L.C.Glaser and T.B.Rushing), LN 438 (1975), 310-324.

[KS] Kirby, R.C. and Siebenmann, L.C. *Foundational Essays on Topological Manifolds, Smoothings, and Triangulations*, Annals of Mathematics Study 88, Princeton University Press, Princeton (1977).

[Ko 1986] Kojima, S. Determining knots by branched covers, in *Low-dimensional Topology and Kleinian Groups* (edited by D.B.A.Epstein), London Mathematical Society Lecture Notes Series 112, Cambridge University Press (1986), 193-207.

[Kr 1987] Kropholler, P.H. (letter to J.A.Hillman, 31 March 1987).

[Kw 1986] Kwasik, S. On low dimensional S-cobordisms, Commentarii Math. Helvetici 61 (1986), 415-428.

[LM 1979] Laitinen, E. and Madsen, I.H. Topological classification of $SL_2(F_p)$ space forms, in *Algebraic Topology, Aarhus* 1978 (edited by J.L.Dupont and I.H.Madsen), LN 763 (1979), 235-261.

[La 1974] Laudenbach, F. Topologie de la dimension trois, homotopie et isotopie, Asterisque 12 (1974).

[Le 1965] Levine, J. Unknotting spheres in codimension two, Topology 4 (1965), 9-16.

[Le 1970] Levine, J. An algebraic classification of some knots of codimension two, Commentarii Math. Helvetici 45 (1970), 185-198.

[Le 1977] Levine, J. Knot modules. I., Trans. Amer. Math. Soc. 229 (1977), 1-50.

[Le 1978] Levine, J. Some results on higher dimensional knot groups, in *Knot Theory, Proceedings, Plans-sur-Bex, Switzerland* 1977 (edited by J.C.Hausmann), LN 685 (1978), 243-269.

[LW 1978] Weber, C. Appendix to [Le 1978], *ibid*, 270-273.

[L1 1986] Lien, M. Construction of high dimensional knot groups from classical knot groups, Trans. Amer. Math. Soc. 298 (1986), 713-722.

[Li 1979] Litherland, R.A. Deforming twist-spun knots, Trans. Amer. Math. Soc. 250 (1979), 311-331.

[Li 1981] Litherland, R.A. The second cohomology of the group of a knotted surface, Quarterly J. Math. Oxford 32 (1981), 425-434.

[Li 1985] Litherland, R.A. Symmetries of twist-spun knots, in *Knot Theory and Manifolds* (edited by D.Rolfsen), LN 1144 (1985), 97-107.

[Li 1988] Livingston, C. Indecomposable surfaces in the 4-sphere,

Pacific J. Math. 132 (1988), 371-378.

[Lo 1981] Lomonaco, S.J., Jr. The homotopy groups of knots I. How to compute the algebraic 2-type, Pacific J. Math. 95 (1981), 349-390.

[Ly 1950] Lyndon, R.C. Cohomology theory of groups with a single defining relation, Ann. Math. 52 (1950), 650-665.

[MW 1950] Mac Lane, S. and Whitehead, J.H.C. On the 3-type of a complex, Proc. Nat. Acad. Sci. 36 (1950), 55-60.

[MKS] Magnus, W., Karrass, A. and Solitar, D. *Combinatorial Group Theory*, (revised edition), Dover Publications, New York (1976).

[Ma 1962] Mazur, B. Symmetric homology spheres, Illinois J. Math. 6 (1962), 245-250.

[McC] McCleary, J. *User's Guide to Spectral Sequences*, Publish or Perish, Inc., Wilmington (1985).

[MS 1985] Meeks, W.H.,III, and Scott, P. Finite group actions on 3-manifolds, Inventiones Math. 86 (1986), 287-346.

[Mi 1987] Mihalik, M.L. Solvable groups that are simply connected at ∞, Math. Z. 195 (1987), 79-87.

[M] Milnor, J. *Introduction to Algebraic K-Theory*, Annals of Mathematics Study 72, Princeton University Press, Princeton (1971).

[Mi 1968] Milnor, J.W. Infinite cyclic coverings, in *Conference on the Topology of Manifolds* (edited by J.G.Hocking), Prindle, Weber and Schmidt, Boston-London-Sydney (1968), 115-133.

[Mi 1975] Milnor, J. On the 3-dimensional Brieskorn manifolds $M(p,q,r)$, in *Knots, Groups and 3-Manifolds - Papers Dedicated to the Memory of R.H.Fox* (edited by L.P.Neuwirth), Annals of Mathematics Study 84, Princeton University Press, Princeton (1975), 175-225.

[Mi 1986] Miyazaki, K. On the relationship among unknotting number, knotting genus and Alexander invariant for 2-knots, Kobe J. Math. 3 (1986), 77-85.

[Mo 1973] Montesinos, J.M. Variedades de Seifert que son recubricadores ciclicos ramificados de dos hojas, Bol. Soc. Mat. Mexicana 18 (1973), 1-32.

[Mo 1983] Montesinos, J.M. On twins in the 4-sphere I, Quarterly J. Math. 34 (1983), 171-199.

[Mo 1984] Montesinos, J.M. On twins in the 4-sphere II,

Quarterly J. Math. 35 (1984), 73-83.

[Mo 1986] Montesinos, J.M. A note on twist spun knots, Proc. Amer. Math. Soc. 98 (1986), 180-184.

[Mo 1968] Mostow, G.D. Quasi-conformal mappings in n-space and the rigidity of hyperbolic space forms, Publ. Math. I.H.E.S. 34 (1968), 53-104.

[Mu 1965] Murasugi, K. On the center of the group of a link, Proc. Amer. Math. Soc. 16 (1965), 1052-1057. Corrigendum *ibid* 18 (1967), 1142.

[N] Neuwirth, L.P. *Knot Groups*, Annals of Mathematics Study 56, Princeton University Press, Princeton (1965).

[New] Newman, M. *Integral Matrices*, Academic Press, New York-London (1972).

[Pa 1978] Pao, P. Non-linear circle actions on the 4-sphere and twisting spun knots, Topology 17 (1978), 291-296.

[Pa 1957] Papakyriakopoulos, C.D. On Dehn's lemma and the asphericity of knots, Ann. Math. 66 (1957), 1-26.

[P] Passman, D.S. *The Algebraic Structure of Group Rings*, John Wiley and Sons Inc., New York-London-Sydney-Toronto (1977).

[Pl 1983] Plotnick, S.P. The homotopy type of four-dimensional knot complements, Math. Z. 183 (1983), 447-471.

[Pl 1983'] Plotnick, S.P. Infinitely many disk knots with the same exterior, Math. Proc. Cambridge Philos. Soc. 93 (1983), 67-72.

[Pl 1984] Plotnick, S.P. Finite group actions and non-separating 2-spheres, Proc. Amer. Math. Soc. 90 (1984), 430-432.

[Pl 1984'] Plotnick, S.P. Fibred knots in S^4 - twisting, spinning, rolling, surgery, and branching, in [GK], 437-359.

[Pl 1986] Plotnick, S.P. Equivariant intersection forms, knots in S^4, and rotations in 2-spheres, Trans. Amer. Math. Soc. 296 (1986), 543-575.

[PS 1985] Plotnick, S.P. and Suciu, A.I. k-invariants of knotted 2-spheres, Commentarii Math. Helvetici 60 (1985), 54-84.

[PS 1987] Plotnick, S.P. and Suciu, A.I. Fibered knots and spherical space forms, J. London Math. Soc. 35 (1987), 514-526.

[Po 1974] Poenaru, V. A note on the generators for the fundamental group of the complement of a submanifold of codimension 2,

Topology 10 (1971), 47–52.

[Qu 1982] Quinn, F. Ends of maps. III: Dimensions 4 and 5, J. Diff. Geometry 17 (1982), 503–521.

[Qu 1984] Quinn, F. Smooth structures on 4-manifolds, in [GK], 473–479.

[Ra 1960] Rapaport, E.S. On the commutator subgroup of a knot group, Ann. Math. 71 (1960), 157–162.

[Ra 1983] Ratcliffe, J.G. A fibered knot in a homology 3-sphere whose group is nonclassical, in *Low Dimensional Topology*, (edited by S.J.Lomonaco, Jr), CONM 20, American Mathematical Society, Providence (1983), 327–339.

[R] Robinson, D.S. *A Course in the Theory of Groups*, Graduate Texts in Mathematics 80, Springer-Verlag, Berlin-Heidelberg-New York (1982).

[Ro 1984] Rosset, S. A vanishing theorem for Euler characteristics, Math. Z. 185 (1984), 211–215.

[RS] Rourke, C.P. and Sanderson, B.J. *Introduction to Piecewise Linear Topology*, Ergebnisse der Mathematik Bd. 69, Springer-Verlag, Berlin-Heidelberg-New York (1972).

[Ru 1987] Ruberman, D. Seifert surfaces of knots in S^4, preprint, Brandeis University (1987).

[Ru 1988] Ruberman, D. The Casson-Gordon invariants in high dimensional knot theory, Trans. Amer. Math. Soc. 306 (1988), 579–595.

[Ru 1986] Rubinstein, J.H. seminar talk, Canberra (July 1986).

[Sa 1981] Sakuma, M. Periods of composite links, Math. Sem. Notes Kobe University 9 (1981), 445–452.

[Sch 1949] Schubert, H. Die eindeutige Zerlegbarkeit eines Knoten in Primknoten, Sitzungsber. Akad. Wiss. Heidelberg, math.-nat. Kl. (1949), 57–104.

[Sch 1953] Schubert, H. Knoten und Vollringe, Acta Math. 90 (1953), 131–286.

[Sc 1983] Scott, P. There are no fake Seifert fibre spaces with infinite π_1, Ann. Math. 117 (1983), 35–70.

[Sc 1983'] Scott, P. The geometries of 3-manifolds, Bull. London Math. Soc. 15 (1983), 401–487.

[Sc 1985] Scott, P. Homotopy implies isotopy for some Seifert fibre spaces, Topology 24 (1985), 341–351.

[Se] Serre, J.P. *Linear Representations of Finite Groups*, Graduate Texts in Mathematics 42, Springer-Verlag, Berlin-Heidelberg-New York (1977).

[Sh 1968] Shaneson, J.L. Embeddings with codimension two of spheres in spheres and h-cobordisms of $S^1 \times S^3$, Bull. Amer. Math. Soc. 74 (1968), 972-974.

[Si 1980] Simon, J. Wirtinger approximations and the knot groups of F^n in S^{n+2}, Pacific J. Math. 90 (1980), 177-190.

[Sp] Spanier, E.H. *Algebraic Topology*, McGraw-Hill, New York (1966).

[Sp 1949] Specker, E. Die erste Cohomologiegruppe von Uberlagerungen und Homotopie-Eigenschaften von dreidimensionaler Mannigfaltigkeiten, Commentarii Math. Helvetici 23 (1949), 303-333

[St 1962] Stallings, J.R. On fibering certain 3-manifolds, in [F], 95-100.

[St 1963] Stallings, J.R. On topologically unknotted spheres, Ann. Math. 77 (1963), 490-503.

[St] Stallings, J. *Group Theory and Three-Dimensional Manifolds* Yale Mathematical Monographs 4, Yale University Press, New Haven-London (1971).

[St 1985] Stark, C.W. Structure sets vanish for certain bundles over Seifert manifolds, Trans. Amer. Math. Soc. 285 (1985), 603-615.

[St 1987] Stark, C.W. L-theory and graphs of free abelian groups, J. Pure Appl. Alg. 47 (1987), 299-309.

[St 1976] Strebel, R. A homological finiteness condition, Math. Z. 151 (1976), 263-275.

[St 1977] Strebel, R. A remark on subgroups of infinite index in Poincaré duality groups, Commentarii Math. Helvetici 52 (1977), 317-324.

[Su 1985] Suciu, A.I. Infinitely many ribbon knots with the same fundamental group, Math. Proc. Cambridge Philos. Soc. 98 (1985), 481-492.

[Su 1971] Sumners, D.W. Invertible knot cobordisms, Commentarii Math. Helvetici 46 (1971), 240-256.

[Su 1976] Suzuki, S. Knotting problems of 2-spheres in the 4-sphere, Math. Sem. Notes Kobe University 4 (1976), 241-371.

[Sw 1974] Swarup, G.A. On a theorem of C.B.Thomas, J. London Math. Soc. 8 (1974), 13-21.

[Sw 1976] Swarup, G.A. An unknotting criterion, J. Pure Appl. Alg. 6 (1975), 291-296.

[Sw 1977] Swarup, G.A. Relative version of a theorem of Stallings, J. Pure Appl. Alg. 11 (1977), 75-82.

[Te 1988] Teragaito, M. Fibred 2-knots and lens spaces, preprint, Kobe University (1988).

[Th 1982] Thurston, W.P. Three dimensional manifolds, Kleinian groups and hyperbolic geometry, Bull. Amer. Math. Soc. 6 (1982), 357-381.

[Tr 1986] Trace, B. A note concerning Seifert manifolds for 2-knots, Math. Proc. Cambridge Philos. Soc. 100 (1986), 113-116.

[Tr 1974] Trotter, H.F. Torsion free metabelian groups with infinite cyclic quotient groups, in *The Theory of Groups* (edited by M.F.Newman), LN 372 (1974), 655-666.

[Tu 1981] Tura'ev, V.G. Three-dimensional Poincaré complexes: classification and splitting, Soviet Math. Doklady 23 (1981), 312-314.

[Wa 1967] Waldhausen, F. Gruppen mit Zentrum und 3-dimensionaler Mannigfaltigkeiten, Topology 6 (1967), 505-517.

[Wa 1968] Waldhausen, F. On irreducible 3-manifolds which are sufficiently large, Ann. Math. 87 (1968), 56-88.

[Wa 1978] Waldhausen, F. Algebraic K-theory of generalized free products I, II, Ann. Math. 108 (1978), 135-256.

[Wa 1967] Wall, C.T.C. Poincaré complexes I, Ann. Math. 86 (1967), 213-245.

[W] Wall, C.T.C. *Surgery on Compact Manifolds*, Academic Press, New York-London (1970).

[W'] Wall, C.T.C. (editor) *Homological Group Theory*, London Mathematical Society Lecture Notes Series 36, Cambridge University Press, Cambridge (1979).

[Wa 1986] Wall, C.T.C. Geometric structures on compact complex analytic surfaces, Topology 25 (1986), 119-153.

[We 1983] Weinberger, S. The Novikov conjecture and low-dimensional topology, Commentarii Math. Helvetici 58 (1983), 355-364.

[We 1987] Weinberger, S. On fibering four- and five-manifolds, Israel J. Math. 59 (1987), 1-7.

[Wh 1950] Whitehead, J.H.C. A certain exact sequence, Ann. Math. 52 (1950), 51-110.

[Wo] Wolf, J.A. *Spaces of Constant Curvature*, (fifth edition), Publish or Perish Inc., Wilmington (1984).

[Ya 1969] Yajima, T. On a characterization of knot groups of some spheres in R^4, Osaka J. Math. 6 (1969), 435-446.

[Ya 1969] Yanagawa, T. On ribbon 2-knots - the 3-manifold bounded by the 2-knots, Osaka J. Math. 6 (1969), 447-464.

[Ya 1977] Yanagawa, T. On cross sections of higher dimensional ribbon knots, Math. Sem. Notes Kobe University 7 (1977), 609-628.

[Yo 1980] Yoshikawa, K. On 2-knot groups with the finite commutator subgroup, Math. Sem. Notes Kobe University 8 (1980), 321-330.

[Yo 1982] Yoshikawa, K. On a 2-knot group with nontrivial center, Bull. Australian Math. Soc. 25 (1982), 321-326.

[Yo 1982'] Yoshikawa, K. A note on Levine's condition for knot groups, Math. Sem. Notes Kobe University 10 (1982), 633-636.

[Yo 1984] Yoshikawa, K. On 2-knot groups with abelian commutator subgroups, Proc. Amer. Math. Soc. 92 (1984), 305-310.

[Yo 1986] Yoshikawa, K. Knot groups whose bases are abelian, J. Pure Appl. Alg. 40 (1986), 321-335.

[Yo 1988] Yoshikawa, K. A ribbon knot group which has no free base, Proc. Amer. Math. Soc. 102 (1988), 1065-1070.

[Ze 1965] Zeeman, E.C. Twisting spun knots, Trans. Amer. Math. Soc. 115 (1965), 471-495.

[Z] Zieschang, H. *Finite Groups of Mapping Classes of Surfaces*, LN 875 (1981).

[ZVC] Zieschang, H., Vogt, E. and Coldewey, H.-D. *Surfaces and Planar Discontinuous Groups*, LN 835 (1980).

[Zi 1979] Zimmermann, B. Periodische Homöomorphismen Seifertscher Faserräume, Math. Z. 166 (1979), 289-297.

[Zi 1982] Zimmermann, B. Das Nielsensche Realisierungsproblem für hinreichend grosse 3-Mannigfaltigkeiten, Math. Z. 180 (1982), 349-359.

[Zi 1986] Zimmermann, B. Finite group actions on Haken 3-manifolds, Quarterly J. Math. Oxford 37 (1986), 499-511.

INDEX